The
PRINCETON FIELD GUIDE *to*
SAUROPOD AND PROSAUROPOD DINOSAURS

The
PRINCETON FIELD GUIDE *to* SAUROPOD AND PROSAUROPOD DINOSAURS

GREGORY S. PAUL

Princeton University Press

Princeton and Oxford

Princeton University Press is committed to the protection of copyright and the intellectual property our authors entrust to us. Copyright promotes the progress and integrity of knowledge created by humans. By engaging with an authorized copy of this work, you are supporting creators and the global exchange of ideas. As this work is protected by copyright, any reproduction or distribution of it in any form for any purpose requires permission; permission requests should be sent to permissions@press.princeton.edu. Ingestion of any IP for any AI purposes is strictly prohibited.

Published by Princeton University Press
41 William Street, Princeton, New Jersey 08540
99 Banbury Road, Oxford OX2 6JX

press.princeton.edu

GPSR Authorized Representative: Easy Access System Europe - Mustamäe tee 50, 10621 Tallinn, Estonia, gpsr.requests@easproject.com

All Rights Reserved
  ISBN 978-0-691-26865-1
  ISBN (e-book) 978-0-691-26872-9

British Library Cataloging-in-Publication Data is available

Editorial: Robert Kirk and Megan Michel Mendonça
Production Editorial: Kathleen Cioffi
Text Design: D & N Publishing, Wiltshire, UK
Jacket Design: Ben Higgins
Production: Steven Sears
Publicity: Matthew Taylor and Caitlyn Robson-Iszatt
Copyeditor: Charles J. Hagner

This book has been composed in LTC Goudy Oldstyle Pro (main text) and Acumin Pro (headings and captions)

Drawing on previous page: *Brontosaurus louisae*

Printed in China

10 9 8 7 6 5 4 3 2 1

CONTENTS

PREFACE

If I were, at about age twenty as a budding paleozoologist and paleoartist, handed a copy of this book by a mysterious time traveler, I would have been shocked as well as delighted. The pages would have revealed a world of new sauropod and related dinosaurs and ideas that I barely had a hint of or had no idea existed at all. My head would have spun at the revelation of the neck spines of *Amargasaurus* and of sauropods so colossal that they not only gave the giant baleen whales a run for their money in terms of sheer bulk but could feed over six stories high in tree crowns—and took only a few decades to get to those sizes! There were sauropodomorphs dwelling in winter chills. I would have noted the new names for some old dinosaurs, including my favorite, *Giraffatitan*, and that old *Brontosaurus* was back! Even *Plateosaurus* would have undergone a remake. There would be the prosauropod and sauropod bearing beds with the familiar yet often-exotic names: Tendaguru, Morrison, Lameta, Kota, Portland, Nemegt, Forest Marble, Navajo Sandstone, Grès a Reptiles, Elliot, Hutton, Paluxy, Alamo, Trossingen, North Horn, Malmros Klint, Lufeng, Arundel. Plus, there would be the novel formations, at least to my eyes and ears: Tiourarén, Las Leoneras, Anacleto, Shishugou, Huincul, Ischigualasto, Hanson, Jinhua, Los Colorados, Pari Aike, Yixian, Allaru, Pebbly Arkose, Villar del Arzobispo, Plottier, Bahariya, Cañadón, Asfalto Tegana, Aïn el Guettar, Suining, Castrillo de la Reina, Lohan Lura, Caturrita, Kallamedu, Cerro Barcino, Lourinha, Portezuelo, Kitadani, Zhanghe, Xert, Maevarano. The sheer number of new dinosaurs would have demonstrated that an explosion in sauropodomorph discoveries and research, far beyond anything that had previously occurred and often based on new high technologies, would mark the end of the twentieth century going into the twenty-first.

A paradigm shift already under way in the late 1960s and especially the 1970s would be confirmed. According to the new view, dinosaurs were not so much reptiles as they were near birds that often paralleled mammals in form and function. Dinosaurs were still widely seen as living in tropical swamps, but we would learn that some lived through polar winters so dark and bitterly cold that low-energy reptiles could not survive. Imagine a small dinosaur shaking the snow off its hairy body insulation while the flakes melted on the scaly skin of a nearby titanic dinosaur whose body, oxygenated by a birdlike respiratory complex and powered by a high-pressure four-chambered heart, produced the heat needed to prevent frostbite.

Producing this volume has been satisfying in that it has given me yet more reason to more fully achieve a long-term goal: to illustrate the skeletons of almost all prosauropod and sauropod species for which sufficiently complete material is available. These have been used to construct the most extensive library of side-view life studies of these dinosaurs in print to date. The result is a work that covers what is now two centuries of scientific investigation into a group of animals that helped rule the continents for over 160 million years. Enjoy the travel back in time.

ACKNOWLEDGMENTS

A complaint earlier this century on the online Dinosaur List by Ian Paulsen about the absence of a high-quality dinosaur field guide led to the production of a first and then a second edition of *The Princeton Field Guide to Dinosaurs*. The exceptional success of those editions, combined with the continuous flux of new discoveries and research, led not only to the production of a third edition but also to this field guide. Many thanks to those who provided the assistance over the years that made these books possible, including John McIntosh, Peter Galton, Kenneth Carpenter, Asier Larramendi, Philip Currie, Robert Bakker, Frank Boothman, Michael Brett-Surman, Daniel Chure, Kristina Curry Rogers, Steven and Sylvia Czerkas, James Farlow, Hermann Jaeger, Rodolfo Coria, Mike Fredericks, Donald Glut, Mark Hallett, Nicholas Hotton, Guy Leahy, Charles Martin, Fernando Novas, Armand Ricqles, Masahiro Tanimoto, Michael Taylor, Matthew Lamanna, Boris Sorkin, Robert Telleria, Matthew Wedel, Lawrence Witmer, Ben Creisler, Dan Varner, Ralph Molnar, Tracey Ford, Skye McDavid, Jens Lallensack, Saswati Bandyopadhyay, Ji Shuan, Clint Boyd, Tyler Greenfield, Mikko Haaramo, John Schneiderman, Michael Malagold, Peter Moon, John Jackson, and many others. I would also like to thank all those who worked on this book for Princeton University Press: Robert Kirk, Megan Mendonça, Kathleen Cioffi, Charles J. Hagner, Namrita and David Price-Goodfellow, Ben Higgins, and Steven Sears.

Brontosaurus

The great sauropod *Giraffatitan*

INTRODUCTION

HISTORY OF DISCOVERY AND RESEARCH

Dinosaur remains have been found by humans for millennia. In the West, the claim in the Genesis creation story that the planet and all life were formed just 2,000 years before the pyramids were built hindered the scientific study of fossils. By the early 1800s, the growing geological evidence that Earth's history was much more complex and extended back into deep time began to free researchers to consider the possibility that long-extinct and exotic animals once walked the globe.

Modern dinosaur paleontology began in the 1820s in southern England with the finding of teeth and fragmentary bones of predatory *Megalosaurus* and the ornithischian *Iguanodon*. Also presented at a science meeting in 1825 were a few gigantic bones attributed to an oversized crocodilian or perhaps a whale. Also then going unrecognized for what it was was a British sauropod tooth. That the Jurassic-age Forest Marble sediments the sauropod fossils came from is marine contributed to the confusion. During the 1830s, disarticulated remains of a small Late Triassic prosauropod were named *Thecodontosaurus*. In 1841, Richard Owen named the Forest Marble material *Cetiosaurus* and the tooth *Cardiodon*, still thinking they represented some form of seagoing reptile. Next year, Owen coined the term "Dinosauria" to accommodate the growing collection of peculiar fossil bones, although he included neither *Thecodontosaurus* nor *Cetiosaurus* in the group. Owen had pre-evolutionary concepts of the development of life, and he envisioned dinosaurs as elephantine versions of reptiles, so they were all restored as heavy-limbed quadrupeds. This led to the first full-size dinosaur sculptures for the grounds of the Crystal Palace in the 1850s. These helped initiate the first wave of dinomania, as they excited the public, but

titanic *Cetiosaurus* was not among them—perhaps it was too large for the budget. It was omitted even though in 1850 the first sauropod long bone from English Cretaceous beds was named *Pelorosaurus* by the same Gideon Mantell who had published *Iguanodon* a quarter century prior. The elongated humerus, bigger than that of any elephants, left no doubt that it belonged to a land walker. Also found was a rare patch of sauropod skin. A British mini bone war was under way, with participants sniping at one another over priorities and interpretations of the frustratingly incomplete remains of the biggest of the terrible reptiles. Meanwhile, on the continent, the first prosauropod fossils were named *Plateosaurus* in the 1830s, having been found in German Triassic sediments. Down in South Africa were found substantial prosauropod fossils of *Massospondylus*. Next decade saw the discovery of large sauropod eggs in southern France, although they would not be realized as such for decades.

The first complete dinosaur skeletons, uncovered in Europe shortly before the American Civil War, were those of small examples, the armored ornithischian *Scelidosaurus* and the birdlike theropod *Compsognathus*. The modest size of these fossils limited the excitement they generated among the public. Found shortly afterward in the same Late Jurassic Solnhofen sediments as *Compsognathus* was the "first bird," *Archaeopteryx*, complete with teeth and feathers. The remarkable mixture of avian and reptilian features preserved in this little dinobird did generate widespread interest, all the more so because the publication of Charles Darwin's theory of evolution at about the same time allowed researchers to put these dinosaurs in a more proper scientific context.

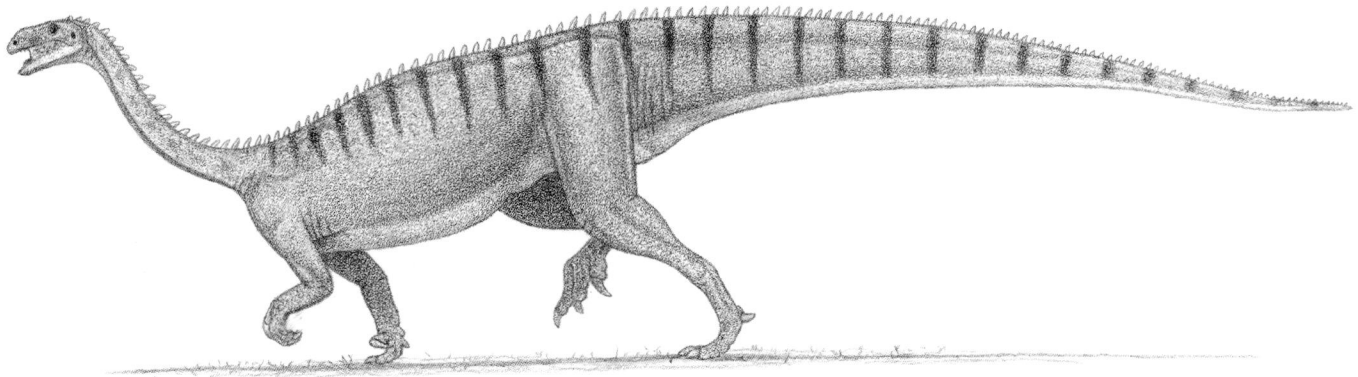

Plateosaurus

At this time, the action was shifting to the United States. Before the Civil War, incomplete dinosaurs had been found on the Eastern Seaboard. But matters really got moving when it was realized that the often forest-free tracts of the West just being cleared of the First Peoples offered hunting grounds that were the best yet for the fossils of extinct titans. This quickly led to the "bone wars" of the 1870s and 1880s, in which Edward Cope and Charles Marsh, having taken a dislike for one another that was as petty as it was intense, engaged in a bitter and productive competition for dinosaur fossils, most spectacular among them sauropods, that would produce an array of quality skeletons, sometimes with skulls. It became possible to appreciate the form of classic Late Jurassic sauropods *Apatosaurus*, *Brontosaurus*, *Barosaurus*, *Diplodocus*, *Camarasaurus*, *Amphicoelias*, *Haplocanthosaurus*, and *Brachiosaurus* from locales named Como Bluff, Garden Park, Bone Cabin, and Sheep Creek. Stretching from southern Canada to New Mexico, and from the front of the Rockies to the middle of the Plains, the uniquely extensive Morrison Formation, with its exceptional sauropodian diversity, remains the premier paleoregional exemplar of the great group. Popular interest in the wonderful beasts was further boosted. For the first time, the bizarro-world shapes of these whale-sized beasts—a cross between elephants, giraffes, and snakes, bearing small heads at the ends of sometimes crazily long necks—was becoming apparent. And the sauropods were themselves diverse in form. Some skulls were short and deep and big-toothed, others long and low, like horses with delicate buckteeth at the front. Necks ranged from moderately to astonishingly long, and tails were the same, with some ending in long whips. Hips were in some cases modest in size or oversized with tall vertebral spines above. Arms could be short and shoulders low; other forelimbs were long and elegant with shoulders set high; and other forelimbs were in between. While the hindfeet were very short, broad, and padded like elephants, they bore banana-shaped claws. The hands were very different, columnar structures, fairly to quite long, unpadded, with thumb claws varying from big to nearly absent. The pneumatic nature of the vertebrae—sophisticated features otherwise limited to birds and pterosaurs—became apparent. Marsh coined the term "Sauropoda," which is neither evocative, nor informative in that sauropods do not have lizard-like feet. Inspired in good measure by the sauropods, Cope devised what became known as Cope's Rule, the propensity for large-bodied animal clades to evolve full-blown gigantism. In the early 1900s, a Morrison apatosaur was tagged *Elosaurus*, a name soon dropped but that is looking valid after all. The finding of what would become Dinosaur National Monument added to the knowledge of the Morrison. Back east, the 1880s saw the discovery of the fairly complete, small, Early Jurassic prosauropod *Anchisaurus* in New England, while outside of Washington, DC, were found rare juvenile sauropod remains of the mid-Cretaceous sauropod *Pleurocoelus*.

Not that the United States alone was the focus of all things sauropodian in the later nineteenth century. In England, enough *Cetiosaurus* bones had shown up to help contribute to restoring the body form of the group. In the jewel of the British crown, the first of the titanosaurs that ruled the sauropod roost in the Late Cretaceous was uncovered in India. In the new century, the sauropod scene shifted to southeastern Africa, where prior to the Great War the

Allosaurus, *Camarasaurus*, *Diplodocus*, and *Stegosaurus*

colonial Germans uncovered at exotic Tendaguru the Late Jurassic supersauropod *Giraffatitan* (was *Brachiosaurus*), as well as the relatively small *Dicraeosaurus*. The scale of the Tendaguru Formation excavation, involving hundreds of local workers, has not been seen before or since. It included more good skull material.

At first, folks had to do with first-generation illustrations of the skeletons. Then, in 1904, a cast replica of a *Diplodocus* skeleton was put on temporary display at an exhibition by Andrew Carnegie, who had paid for the finding of a big dinosaur skeleton to grace his new museum in Pittsburgh. In response to a casual request by the King of England, casts ended up being distributed around the globe, giving publics their first look at a sauropod skeleton in all its glory. A *Brontosaurus* was mounted in 1905 at the American Museum of Natural History in New York City, albeit with the wrong type of skull.

In part because sauropods were just so big and life spent largely on land seemed a structural stretch, they were characterized from the beginning in the main as sluggish, cold-blooded, reptilian equivalents of hippos. They were thought to have spent the bulk of their lives lolling in the body-supporting warm waters of great subtropical looping rivers and placid lakes spread across rain-drenched floodplains, where they foraged with their little heads at the end of snake necks on soft, easily digestible plants covering the bottoms and lining the shores, protected from the depredations of the oddly hydrophobic flesh-eating dinosaurs, and emerged on occasion to lay their eggs. The high-set position of their external nares seemed to support this aquatic theory. Also suggested was eating mussels, though few paleozoologists went for that. That the pneumatic bones were not what was expected in amphibious forms went ignored; hippos have high-density bones. As Carnegie's *Diplodocus* mounts were being distributed across the pond, a European researcher contended that they should not be posed with vertical legs like modern elephants, but with croc-like sprawling limbs, inspiring a vehement response from American colleagues, who noted, among other items, that the inset cylindrical hip socket of dinosaurs did not allow for any other than an erect posture. The paleoartists of the time, led by the renowned Charles Knight, followed the paleobiologists in usually rendering the beasts as dwellers of swamps and marshes, often showing only the top of the body and neck above the surface. What all agreed on was that sauropods, being reptiles, dragged their normally limp tails like crocodilians and lizards when ashore, activating them only to lash out at predators and rivals.

There was some dissent. A young Henry Osborn thought gracile *Diplodocus* to be an active animal able to rear on its hindlimbs, an idea picked up in a one-off illustration by Knight. Cutting-edge analyst Elmer Riggs, influenced by the long, lanky limbs of the *Brachiosaurus* he had found and the powerful skeleton of *Apatosaurus*, understood that sauropods were built in the elephantine mold, not the short-limbed, tubby form of hippos. At the last turn of the century, he proposed sauropods as land-striding tree browsers like similarly long-necked giraffes. He thought that the short-armed, especially big-hipped diplodocids stood on two legs like elephants in circuses and in the wild to feed extra high. Unfortunately, this cutting-edge science did not gain traction, and the waterbound sauropod remained the conservative standard for another lifetime. Dinosaurology as a whole became rather ossified, like the bones under study, with the giant beasts most of all widely portrayed as slothful, dim-witted evolutionary dead ends doomed to extinction. They were considered examples of the "racial senescence" theory that was widely held among researchers who preferred a progressive concept of evolution at odds with the more random reality of Darwinian natural selection. That birds were seen as a remote offshoot of archosaurs having nothing to do with clunky dinosaurs did not help matters. Meanwhile, the *Star Wars–Jurassic Park* of its time, RKO's *King Kong* of 1933, amazed audiences with its dinosaurs seemingly brought to life, including a swamp *Brontosaurus* that had inordinate anger issues with rafting humans. Two major film comedies, 1938's *Bringing Up Baby*, starring Cary Grant and Katherine Hepburn, and 1949's *On the Town*, featuring Gene Kelly and Frank Sinatra, included climactic scenes in which sauropod skeletons at a semifictional New York museum collapse because of the ill-advised actions of leading characters. Yet the very popularity of dinosaurs gave them a circus air that convinced many scientists that they were beneath their erudite scientific dignity and attention. At the Los Angeles museum, the paleomammologists even ended up trashing—some of it seems to have ended up as fill for a driveway—most of a *Brontosaurus* skeleton they found unsuitable for putting on display to draw in the crowds. Best to focus on the mammals of the Cenozoic, which included among them the ancestors of the humans who studied the prehistoric beasts.

Meanwhile, in Germany, increasing numbers of *Plateosaurus* skeletons, many dozens and counting, were being excavated, showing what the medium-sized prosauropods of the Late Triassic and Early Jurassic were like: small, lightly built, at least partly bipedal, possessing narrow heads at the end of fairly long, slender necks, with long, bulky bodies, moderately long but deep tails, and short, flexed arms bearing big thumb claws. Their distinctive skeletons began to grace museums around the planet. Friedrich von Huene invented Prosauropoda to cover the early Sauropodomorpha he also devised to include the ancestors of sauropods. In Berlin, the remains of *Giraffatitan*, the greatest sauropod of the day, towering at four stories, were put up only to be shortly disassembled to save them and other dinosaur mounts from Allied bombing.

Between the wars in Arizona and later Utah, a few fossils tagged *Alamosaurus* were uncovered by Smithsonian crews, showing that the great group had made it to the end of the dinosaur era, as other plant-eating dinosaurs evolved an array of new forms—armored, horned, and duck-billed. So did Romanian *Magyarosaurus*, as well as South American *Laplatasaurus*. Also out of South America were some very large titanosaur bones that exceeded the brachiosaurs. Down under, Jurassic *Rhoetosaurus* and Cretaceous *Austrosaurus* showed up. Egypt produced similarly incomplete mid-Cretaceous *Aegyptosaurus*, Morocco *Rebbachisaurus*. Despite poor national conditions in China, foreign and local paleontologists announced remains of the Middle and Late Jurassic sauropods *Omeisaurus* and *Euhelopus*, the latter with a rare skull, as well as skulls and skeletons of prosauropods as per Early Jurassic *Lufengosaurus* and *Yunnanosaurus*. Earlier *Melanorosaurus* was from South America. A bit of an advance was made in mounting sauropods' skeletons: the original *Diplodocus* mounts had the tails dropping down and back immediately from the hips in line with the supposed reptilian drooping posture. But this left the first tail bones well out of articulation. This was realized, and at the Smithsonian and Denver museums the tail-base vertebrae were properly lined up to emerge horizontal from the pelvis. But then the tail arced down to the ground reptilian style.

At the end of the 1930s, working for the American Museum of Natural History, Roland Bird revealed the trackways of sauropods from the mid-Cretaceous of Texas. Found in modern streambeds, the then coastal carbonate flats recorded the comings and goings of beasts from 50 tonne adults down to 1 tonne juveniles—but not really small youngsters. The footprints confirmed that the legs were under the body, not sprawled outward. These trace fossils were data goldmines that went largely ignored. Some of the Texas trackways were clearly made at the same moment by large adult/juvenile herds moving as one. This did not stop dinosaurs from being seen as small-brained individualists. Also of note was that there were no tail dragmarks. This was explained away as due to the dinosaurs wading in water that buoyed their tails, even though there was no evidence for the water, and very few other dinosaur trackways had tail dragmarks either.

Indeed, after World War II, sauropodology remained in the doldrums scientific and artistic, with the likes of Rudolph Zallinger and Zdenek Burian continuing to portray sauropods as bound to rivers and lakes. One of the few researchers to show a deep interest in the beasts was nuclear particle physicist John (Jack) McIntosh. The *Giraffatitan*, *Dicraeosaurus*, and *Plateosaurus* were remounted in Communist East Berlin and described in detail. And there was fieldwork going on, especially in China, where, despite a revolution gone bad, prosauropods and sauropods were being excavated and described. Also outstanding was how the Poles ventured to Mongolia in the 1960s, discovering the skull and a skeleton of Late Cretaceous sauropods.

The tired, unscientific mythology of reptilian, water-loving sauropods began to break apart in the late 1960s and 1970s as part of the great Dinosaur Renaissance, which radically transformed the field. John Ostrom pointed out that, like the legs of other dinosaurs, those of sauropods were built on the mammalian pattern, not reptilian, so it could not be assumed that sauropods were cold-blooded—although he remained fairly conservative, suggesting that sauropods maintained steady temperatures via bulk. Ostrom student Robert Bakker took things further to their logical paleozoological conclusion, contending that all dinosaurs constituted a distinct group of archosaurs whose biology and energetics were more avian than reptilian. And he dragged sauropods out of the water onto the shore, reviving forever the Riggs view of terrestrial sauropods. A research team showed that the Morrison floodplain was largely seasonally arid land with open conifer cycad woodlands cut by braided rivers too shallow to immerse the huge beasts—back in the 1950s, it had been observed that sauropods could not have breathed in deep rivers and lakes anyway, due to the high pressure of the water. Bakker startlingly illustrated the land concept with a pair of *Barosaurus* striding like elephant/giraffes across the Morrison floodplain, tails clear of the ground as per the trackways, necks held erect to view the landscape. The air-filled vertebrae were seen as part of a birdlike respiratory complex that helped oxygenate high metabolic rates and food budgets like those of similarly gigantic mammals. The long necks allowed the land titans to feed on the crowns of trees to acquire the very high energy intake needed to be tall and ponderous while living in the planetary gravity well. Skeletal drawings revived rearing as a way to feed as high as possible. The new view of energetic, social, high-browsing sauropods would be as initially hotly contested as it has become the modern consensus.

As interest in sauropods as objects of scientific research and rising popular interest picked up, so did the curiosity of those who were getting new bones out of the ground. Jim (Dinosaur) Jensen opened a Colorado Morrison quarry that started producing record breakers, including appropriately named *Supersaurus*. The Chinese continued to churn out assorted sauropodomorphs, including extra long-necked mamenchisaurs and early shunosaurs. Also early in the sauropod phylogeny were the Moroccan *Atlasaurus* with its extra long limbs, especially arms, Indian *Barapasaurus*, and southern African *Vulcanodon*.

At the same time, researchers from outside paleontology stepped into the field and built up the evidence that the impact of a mountain-sized asteroid was the long-sought great dinosaur killer. This extremely controversial and contentious idea turned into the modern paradigm when a state-sized meteorite crater was found in southeastern Mexico dating to the end of the dinosaur era. At the same time, the volcanic Deccan Traps of India were proposed as an adjunct in the demise of the sauropods and all other dinosaurs, aside from a few birds.

These radical and controversial concepts greatly boosted popular attention to dinosaurs, culminating in the *Jurassic Park* novels and films, which sent dinomania to unprecedented heights. The first dinosaur seen in *Jurassic Park* by the stunned paleontologists was an athletic brachiosaur rearing up to chomp on a tree crown—albeit not anatomically well executed. The elevated public awareness was fueled by digital technology in the form of touring exhibits of robotic dinosaurs, sans full-scale sauropods, since they were too big. This time, the interest of paleontologists was raised as well, inspiring the second and ongoing golden age of dinosaur discovery and research, which is surpassing that which has gone before. Assisting the work are improved scientific techniques in the area of evolution and phylogenetics, including cladistic genealogical analysis, which has helped improve the investigation of dinosaur relationships. New generations of artists have portrayed prosauropods and sauropods with the "new look" that lifts tails in the air and gets feet off the ground to represent the more dynamic gaits that are in line with the more active, often-social lifestyles the researchers now favor. The painting on page 130 of a *Diplodocus* rearing up to fend off *Allosaurus* while a juvenile flees underneath inspired the American Museum of Natural History to mount skeletons of *Barosaurus* of the same ages along with that of the attacking predatory dinosaur in the great entry hall. While much of the reptile was being taken out of the sauropods, paleosculptor Stephen Czerkas put some back by showing that they were adorned with midline spike frills, causing fellow paleoartists to scramble to apply them to their suddenly outdated restorations. Meanwhile, Jack McIntosh showed that mounted skeletons of the robust apatosaurs were being wrongly fitted with camarasaur-like skulls when they actually shared the gracile horse heads of their close diplodocid relations.

Prosauropods and sauropods are being found and named at an unprecedented rate, with extensive efforts under way on all continents. Especially important has been the development of local expertise made possible by the rising economies of many nations, reducing the need to import expertise.

In South America, Argentine and American paleontologists collaborated in the 1960s and 1970s to reveal the first Middle and Late Triassic protodinosaurs, showing that the very beginnings of dinosaurs predatory and herbivorous started among surprisingly small archosaurs. This put an end to the idea that different dinosaur groups emerged separately from the thecodonts. Since then, *Eoraptor* and *Buriolestes* have helped show what the unpretentious beginnings of sauropodomorphs looked like. Argentina has been the source of endless prosauropod and sauropod remains from the Triassic to the end of the Cretaceous that include the supertitanosaurs *Argentinosaurus* and *Patagotitan*, with an array of names often based on limited material. Among the new taxa, *Amargasaurus* sported spectacular neck spines, and *Brachytrachelopan* sports an even shorter neck. Among the most extraordinary finds from that continent have been titanosaur nesting grounds that allow us to see how the greatest land animals of Earth's history reproduced. And the same goes for prosauropods, with *Mussaurus* being known from eggs with embryos to

adults. China, too, has been churning out sauropodomorphs, from sometimes very great down to small, including nests from the Late Triassic to the Late Cretaceous, leading to a confusion of names that needs sorting out. Other regions of Asia, Japan included, are productive, with Mongolian *Erketu* having the proportionally longest neck to date. And in India, sauropod dung indicates that grasses were appearing toward the end of the Cretaceous period, and bone fragments hint at ultrasized titanosaurs. Toward the opposite end of the sauropod size spectrum, the examples out of Africa include small-bodied *Nigersaurus*, which grazed ground cover with squared-off jaws at the end of a short neck. Europe is also producing more small sauropods, including dwarf examples that lived on the Cretaceous island archipelago that was the Europe of the time. First found in Europe are the turiasaurs, a previously unknown clade of the later Jurassic and earlier Cretaceous popping up on various continents. From the Late Jurassic of Portugal and Spain are coming large camarasaurs, brachiosaurs, and diplodocids very similar to those from the contemporaneous Morrison Formation of the American West, which itself is continuing to be a source of new species—and a reconsideration of an old lost bone now named *Maraapunisaurus* is affirming it was a supergiant. In that land, the Early Cretaceous Cedar Mountain beds are turning up the sauropods *Mierasaurus*, *Moabosaurus*, *Cedarosaurus*, *Abydosaurus*, and *Venenosaurus*. The venerable Morrison Formation continues to be a big source of sauropod fossils, which in association with old fossils are resulting in new and revived names, including the return of the thunder lizard *Brontosaurus*. Remaining is North America's atypical "sauropod gap," in which for reasons unknown the behemoths did not dwell on the continent for most of the long Late Cretaceous until near the end of the dinosaur era, and even then did not move beyond the Southwest. Australia continues to generate incomplete Jurassic and Cretaceous sauropods.

Of special interest are bits of bone that establish that prosauropods dwelled in Antarctica, where the dark Early Jurassic winters were chilly. Also of import is that the transition from archaic prosauropods to the heavier, taller, more elephantine sauropods is becoming known from remains from differing locations, among them Chinese *Yizhousaurus* and *Gongxianosaurus*, and Moroccan *Tazoudasaurus*.

On a global scale, the number of sauropodomorph trackways that have been discovered is in the millions, with the substantial majority being sauropodian. The sheer abundance of footprints is logical in that a given dinosaur could potentially contribute only one skeleton to the fossil record but could make innumerable footprints. In a number of locations, trackways are so abundant that they form what have been called "dinosaur freeways" that were laid down mostly by sauropods as they made their way through a Mesozoic world. A number of the trackways were formed in a manner that suggests their makers were moving in herds. A few may record the attacks of predaceous theropods on the great herbivores.

While sauropods being largely landlubbers is the modern paradigm, a dispute arose over the posture of their necks around the turn of the century. The Bakker drawing of the *Barosaurus* pair showed their necks emerging almost straight up from the shoulders, which is not possible in such diplodocids. And much as it had been shown that deepwater sauropods could not have worked because the water pressure would have prevented breathing, it was contended that the extreme blood pressure needed to pump oxygen to a brain held many meters above heart level precluded the sauropods from raising their heads so high, so their necks must have been subhorizontal. Which eliminates the functional need for long necks. Some researchers agreed, arguing that the articulation of the neck-base vertebrae of all sauropods precluded their necks from being elevated well above shoulder level. The idea of sauropods immersed in the shallows while feeding on shoreline vegetation was briefly revived. But broader examinations showed that most sauropods' neck bases could readily be tilted strongly upward, a point proven by two neck-base vertebrae of an old individual that had fused together at an upcurved angle that would not be possible according to the level-neck hypothesis. And diplodocids were well adapted for rearing on their hindlegs and tails to bring their necks vertical. One way or another, many sauropods could lift their heads many stories high regardless.

The history of prosauropod and sauropod research is not just one of new ideas and new locations; it is also one of new techniques and technologies. The turn of the twenty-first century has seen paleontology go high tech, with the use of ultrafast computations for processing data and high-resolution CT scanners to peer inside fossils without damaging them. Skeletal and life restorations are being generated in 3-D digital format—whether this has improved the process is another matter. Dinosaurology has also gone microscopic and molecular in order to assess the lives of dinosaurs at a more intimate level, telling us how fast they grew, how long they lived, and at what age they started to reproduce. Bone isotopes are being used to help determine dinosaur diets. Meanwhile, the *Jurassic World* franchise helps sustain popular interest in the group even as it presents an obsolete, prefeather image of the birds' closest relations.

The evolution of human understanding of sauropodomorphs has undergone a series of dramatic transformations since they were scientifically discovered almost 200 years ago. This is true because they are a group of "exotic" animals whose biology was not obvious from the start, unlike fossil mammals or lizards. It has taken time to build up the knowledge base needed to resolve their true form and nature. The latest revolution is still young. Dinosaurology has matured in that it is unlikely that a reorganization of similar scale will occur in the future, but we now know enough about the inhabitants of the Mesozoic to have the basics well established. Sauropods will not return to a hippo-like lifestyle, and their tails will not be chronically plowing through Mesozoic muds. Sauropod dinosaurs are no longer so mysterious. Even so, the research is nowhere near its end. To date, about 275 valid prosauropod and sauropod species in over 200 genera have been discovered and named. This probably represents at most a quarter, and perhaps a much smaller fraction, of the species that have been preserved in sediments that can be accessed. And, as astonishingly strange as many of the dinosaurs uncovered so far have been, there are equally odd species waiting to be unearthed. Reams of work based on as-yet-undeveloped technologies and techniques will be required to provide further details about both dinosaur biology and the world in which they lived. And although a radical new view is improbable, there will be many surprises.

WHAT ARE PROSAUROPOD AND SAUROPOD DINOSAURS?

To understand what a dinosaur is, we must first start higher in the scheme of animal classification. The Tetrapoda are the vertebrates adapted for life on land—amphibians, reptiles, mammals, birds, and the like. Amniota comprises those tetrapod groups that reproduce by laying hard-shelled eggs, with the proviso that some have switched to live birth. Among amniotes are two great groups. One is the Synapsida, which includes the archaic pelycosaurs, the more advanced therapsids, and mammals, which are the only surviving synapsids. The other is the Sauropsida, much the same as the Reptilia, all but some early forms of which belong to the great Diapsida, which included most of the ancient marine reptiles that are detailed in the Princeton field guide on that set of creatures. Surviving diapsids include turtles, the lizard-like tuataras, true lizards and snakes, crocodilians, and birds. The Archosauria is the largest and most successful group of diapsids and includes crocodilians and dinosaurs. Birds are literally flying dinosaurs.

Archosaurs also include the basal forms informally known as thecodonts because of their socketed teeth, themselves a diverse group of terrestrial and aquatic forms that include the ancestors of crocodilians and the winged pterosaurs (covered in another Princeton field guide), although those fliers were not intimate relatives of dinosaurs and birds.

Euparkeria basal archosaur

Researchers concur that the dinosaurs were monophyletic in that they shared a common ancestor that made them distinct from all other archosaurs, much as all mammals share a single common ancestor that renders them distinct from all other synapsids. This consensus is fairly recent—before the 1970s, it was widely thought that dinosaurs came in two distinct types that had evolved separately from thecodont stock: the Saurischia and Ornithischia. It was also thought that birds had evolved as yet another group independently from thecodonts. Dinosauromorpha are dinosaurs and close relations that consist of the protodinosaurs. Dinosauriformes exclude the lagerpetids, which

look close to pterosaurs. Dinosauria itself is formally defined as the phylogenetic clade that includes the common ancestor of *Triceratops* and birds and all their descendants.

In anatomical terms, one of the features that most distinguish dinosaurs centers on the hip socket. The head of the femur is a cylinder turned in at a right angle to the shaft of the femur. The head fits into a cylindrical, internally open hip socket. This allows the legs to operate in the nearly vertical plane characteristic of the group, with the feet directly beneath the body. You can see this system the next time you have chicken thighs. The ankle is a simple fore-and-aft hinge joint that also favors a vertical

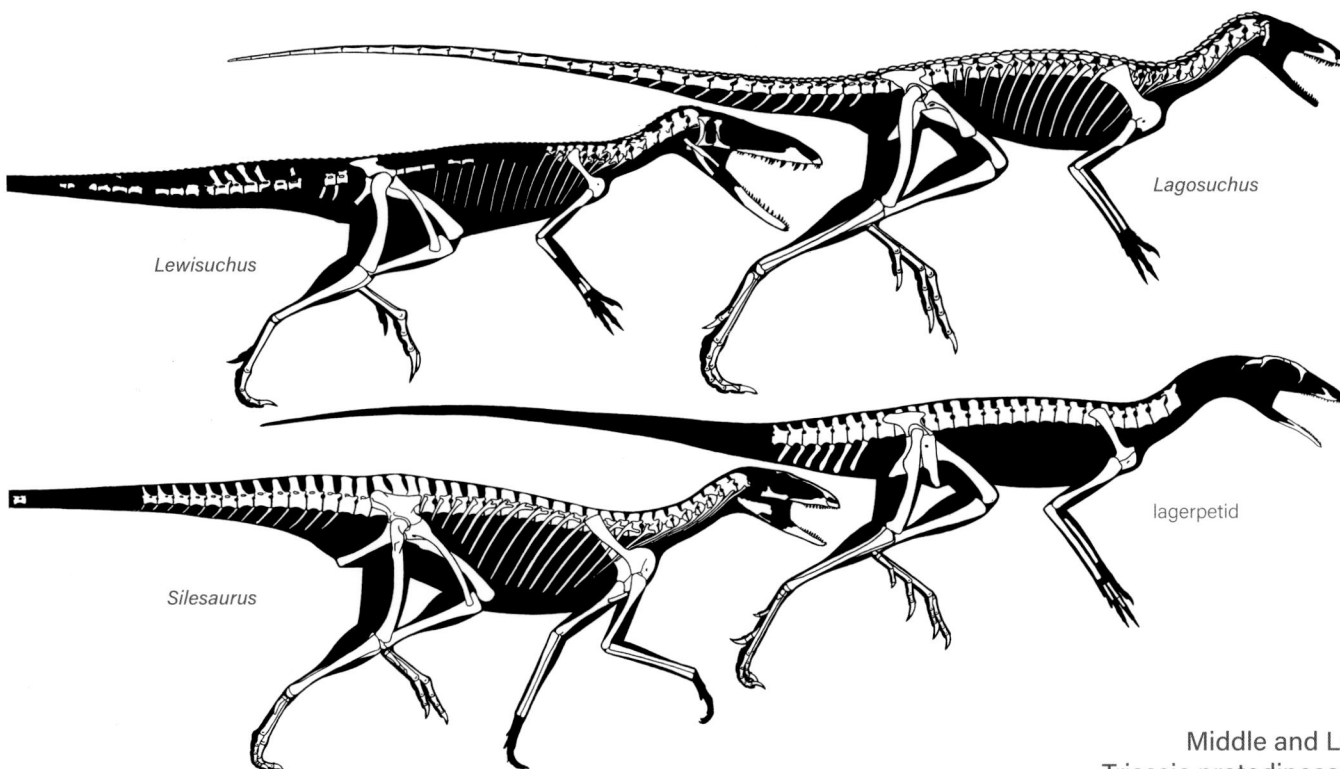

Middle and Late
Triassic protodinosaurs

15

leg posture. Dinosaurs were "hindlimb dominant" in that either they were bipedal or, even when they were quadrupedal and had long arms, most of the animal's weight was borne on the legs, which were always built more strongly than the arms. The hands and feet were generally digitigrade, with the wrist and ankle held clear of the ground. All dinosaurs shared a trait also widespread among archosaurs in general, the presence of large and often remarkably complex sinuses and nasal passages.

Aside from the above basic features, dinosaurs, even when we exclude birds, were an extremely diverse group of animals, rivaling mammals in this regard. Dinosaurs ranged in form from nearly birdlike types, such as the sickle-clawed dromaeosaurs, to rhino-like horned ceratopsians to armor-plated stegosaurs to elephant- and giraffe-like sauropods and dome-headed pachycephalosaurs.

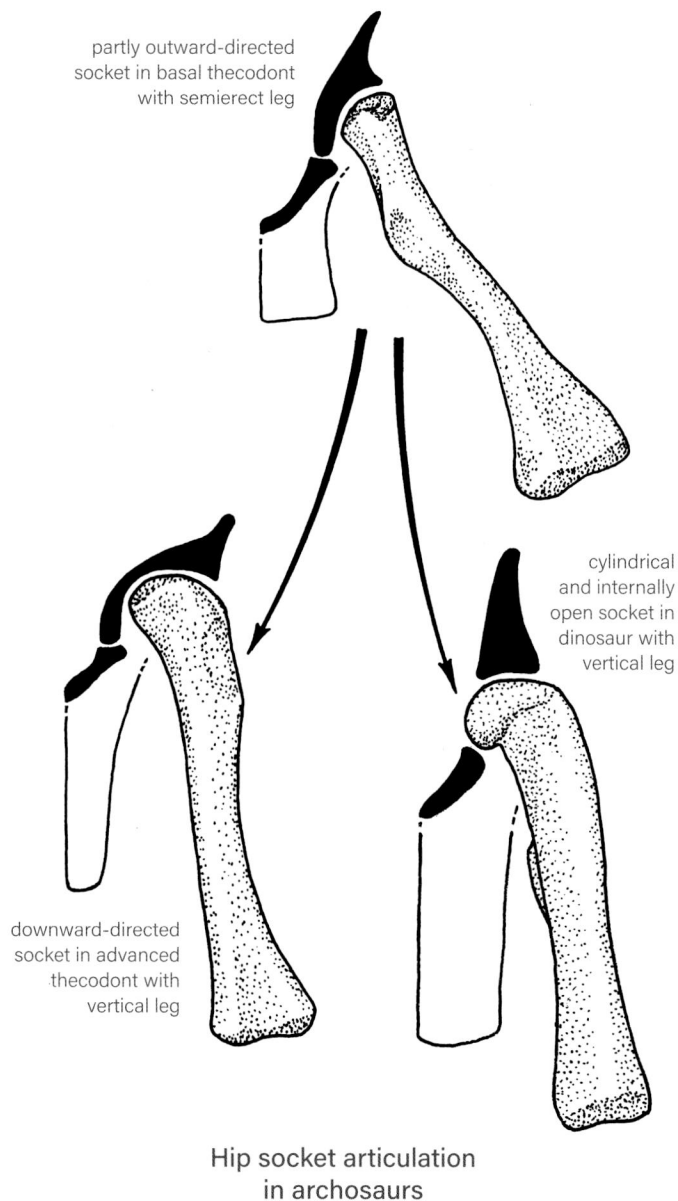

They even took to the skies in the form of birds. However, dinosaurs were limited in that they were persistently terrestrial. Although some dinosaurs may have spent some time feeding in the water like moose or fishing cats, at most a few became strongly amphibious in the manner of hippos, much less marine like seals and whales. The only strongly aquatic dinosaurs are some birds. The occasional statement that there were marine dinosaurs is therefore very incorrect—these creatures of Mesozoic seas were various forms of reptiles that had evolved over the eons. The situation within early dinosaurs is interesting because it has become chaotic. For over a century, dinosaurs have generally been segregated into the two classic groups: the Saurischia, which include the largely herbivorous sauropodomorphs and the usually predaceous theropods, birds included; and the vegetarian Ornithischia, which despite the name have nothing to do with birds. But while ornithischians are clearly a real, monophyletic clade, the saurischian group has always been weak. One of the key features of saurischians is their birdlike air-sac respiratory complexes. But early examples of the sauropodomorphs may lack them, in which case their presence in the two groups is due to convergence, not shared heritage. Another and major part of the problem is that the earliest species of a given group are all pretty similar to one another, and it is difficult to parse out the few, still subtly different characters that show they are at the base of one of the greater, more sophisticated groups but not the other. Add to that the incompleteness of the fossil record, and we are missing most of the information needed to sort the situation out. As a result, some researchers have removed the theropods from the Saurischia and more closely allied them with ornithischians in the Ornithoscelida, in which case those dinosaurs not in the latter can be placed in a Paxdinosauria. But there are problems with that scheme, too, and it is possible that the herbivorous sauropodomorphs and ornithischians form a joint clade, the Ornithischiformes or Phytodinosauria, they being another example of unstable terminology. All possibilities remain similarly viable at this time; the situation may not be resolvable with the data on hand, if ever, and this volume does not take a firm stand on the issue.

Whatever dinosaurs they are most related to, sauropodomorphs are a distinctive group of largely herbivorous examples—some omnivory was involved—that had long rows of at least sometimes blunt teeth and did not have tooth-free beaks, and a nonretroverted pelvic pubis. Prosauropods are an informal group at the base of the greater group that can be defined as sauropodomorphs that are not sauropods. Prosauropods are commonly referred to with the polysyllabic technojargon of basal sauropodomorphs, or the similarly tongue-twisting nonsauropod sauropodomorphs, which define them by what they are not. Triangular-skulled, with broad nasals, often bipedal, long-bodied, heavy-tailed, and with flexed arms and legs, prosauropods could run at modest speeds. Sauropods were a different collection, sans the bulging nasals, with stout

partly outward-directed socket in basal thecodont with semierect leg

cylindrical and internally open socket in dinosaur with vertical leg

downward-directed socket in advanced thecodont with vertical leg

Hip socket articulation
in archosaurs

elephantine bodies and columnar legs that prevented them from running, all generally quadrupedal, normally big if not outright titanic.

Because birds are dinosaurs in the same way that bats are mammals, the dinosaurs aside from birds are sometimes referred to as "nonavian dinosaurs." This usage can become awkward, and in general in this book dinosaurs that are not birds are, with some exceptions, referred to simply as dinosaurs, including sauropodomorphs.

The prosauropods and sauropods seem strange, but that is just because we are mammals biased toward assuming the modern fauna is familiar and normal, and past forms are exotic and alien. Consider that elephants—the living animals most like sauropods—are bizarre creatures, with their combination of big brains, massive limbs, oversized ears, a pair of teeth turned into tusks, and noses elongated into hose-like trunks. And giraffes—the other living type most like sauropods—are also peculiar. Nor were sauropodomorphs part of an evolutionary progression that was necessary to set the stage for mammals culminating in humans. What dinosaurs do show is a parallel world, one in which mammals were permanently subsidiary, and the dinosaurs show what largely diurnal land animals that evolved straight from similarly day-loving ancestors should actually look like. Modern mammals are much more peculiar, having evolved from nocturnal beasts that came into their own only after the entire elimination of nonavian dinosaurs. While dinosaurs dominated the land, small nocturnal mammals were abundant and diverse, as they are in our modern world. If not for the accident events that brought the Age of Dinosaurs to a quick end, dinosaurs would probably still be the global norm.

DATING PROSAUROPODS AND SAUROPODS

How can we know that sauropodomorphs lived in the Mesozoic, first appearing in the Late Triassic over 230 million years ago and then disappearing at the end of the Cretaceous 66 million years ago?

As gravels, sands, and silts are deposited by water and sometimes wind, they build up in sequence atop the previous layer, so the higher in a column of deposits a dinosaur is, the younger it is relative to dinosaurs lower in the sediments. Over time, sediments form distinct stratigraphic beds that are called formations. For example, *Apatosaurus*, *Brontosaurus*, *Diplodocus*, *Barosaurus*, *Haplocanthosaurus*, *Amphicoelias*, *Suuwassea*, *Supersaurus*, and *Dystrophaeus* are found in the Morrison Formation of western North America, which was laid down in the Late Jurassic, from 156 to 150 million years ago. Deposited largely by rivers over an area covering many states in the continental interior, the Morrison Formation is easily distinguished from the marine Sundance Formation lying immediately below as well as from the similarly terrestrial Cedar Mountain Formation above, which contains very different sets of dinosaurs. Formed over millions of years, the Morrison is subdivided into lower (older), middle, and upper (younger) levels. So a fossil found in the Sundance is older than one found in the Morrison, a dinosaur found in the lower Morrison is older than one found in the middle, and a dinosaur from the Cedar Mountain is younger still.

Geological time is divided into a hierarchical set of names. The Mesozoic is an era—preceded by the Paleozoic and followed by the Cenozoic—that contained the three progressively younger periods called the Triassic, Jurassic, and Cretaceous. These are then divided into Early, Middle, and Late, except that the Cretaceous is split only into Early and Late despite being considerably longer than the other two periods (this was not known when the division was made in the 1800s). The periods are further subdivided into stages. The Morrison Formation, for example, began to be deposited during the last part of the Oxfordian and continued through the entire Kimmeridgian. Lasting 40 million years, the Cedar Mountain Formation spans all or part of seven stages of nearly all the Early Cretaceous into the beginning of the Late Cretaceous.

The absolute age of recent fossils can be determined directly by radiocarbon dating. Dependent on the ratios of carbon isotopes, this method works only on bones and other organic fossils going back up to 50,000 years, far short of the dinosaur era. Because it is not possible to date Mesozoic prosauropod and sauropod remains directly, we must instead date the formations that the specific species are found in. This is viable because a given dinosaur species lasted only a few hundred thousand to a few million years.

The primary means of absolutely determining the age of sauropodomorph-bearing formations is radiometric dating. Developed by nuclear scientists, this method exploits the fact that radioactive elements decay in a very precise manner over time. The main nuclear transformations used are uranium to lead, potassium to argon, and one argon isotope to another argon isotope. This system requires the presence of volcanic deposits that initially set the nuclear clock. These deposits are usually in the form of ashfalls, similar to the one deposited by Mount Saint Helens over neighboring states, that leave a distinct layer in the sediments. Assume that one ashfall was deposited 144 million years ago, and another one higher in the sediments 141 million years ago. If a sauropod is found in the deposits in between, then we know that the dinosaur lived between 144 and 141 million years ago. As the technology advances and the geological record is increasingly better known, radiometric dating becomes increasingly precise. The further back in time one goes, the greater the margin of error and the less exactly the sediments can be dated. In many cases, the dating is well set and not likely to change—the end of the Mesozoic happened just a dash over 66 million years ago. In other cases, less favorable geological circumstances leave matters less precisely settled; not long ago the shift from the Oxfordian to the Kimmeridgian stage was calculated to have occurred less than 155 million years ago, but it was just updated to over 157 million, a value that may change again based on further research.

Because it was not possible to date geological deposits accurately when the time divisions were mapped and named in the 1700s and 1800s, they later proved to be very irregular in the amount of time covered by each division; for instance, the Norian stage is about 10 times longer than the Hettangian stage.

Volcanic deposits are often not available, and other methods of dating must be used. Doing so requires biostratigraphic correlation, which can in turn depend in part on the presence of "index fossils." Index fossils are organisms, usually marine invertebrates, that are known to have existed for only geologically brief periods of time, just a few million years at most. Assume a prosauropod or sauropod species is from a formation that lacks datable volcanic deposits. Also assume that the formation grades into marine deposits laid down at the same time near its edge. The marine sediments contain organisms that lasted for only a few million years or less. Somewhere else in the world,

the same species of marine life was deposited in a marine formation that includes volcanic ashfalls that have been radiometrically dated to between 83 and 81 million years. We can then conclude that the sauropod in the first formation is also 83–81 million years old.

A number of sauropodomorph-bearing formations lack both volcanic deposits and marine index fossils. It is often not possible to date the organisms in these deposits. It is only possible to broadly correlate the level of development of the prosauropods or sauropods and other organisms in the formation with faunas and floras in better-dated formations, and this produces only approximate results. This situation is especially common in central Asia. The reliability of dating therefore varies. It can be very close to the actual value in formations that have been well studied and contain volcanic deposits; these can be placed in specific parts of a stage. At the other extreme are those formations that, because they lack the needed age determinants, and/or because they have not been sufficiently well examined, can only be said to date from the early, middle, or late portion of one of the periods, an error that can span well over 10 million years. North America currently has the most robust linkage of the geological time scale with its fossil dinosaurs on Earth.

THE EVOLUTION OF PROSAUROPODS AND SAUROPODS AND THEIR WORLD

Prosauropods and then sauropods appeared in a world that was both ancient and surprisingly recent—it is a matter of perspective. The human view that the Mesozoic was remote in time is an illusion that results from our short life span. A galactic year, the time it takes our solar system to orbit the center of the galaxy, is 200 million years. Only one galactic year ago, the prosauropods had just appeared on planet Earth. When those first sauropodomorphs first emerged, our solar system was already well over four billion years old, and 95 percent of the history of our planet had already passed. A time traveler arriving on Earth when dinosaurs first existed would have found it both comfortingly familiar and marvelously different from our time.

As the moon slowly spirals out from Earth because of tidal drag, the length of each day grows. When prosauropods initially evolved, a day was about 22 hours and 45 minutes long, and the year had about 385 days; when sauropods went extinct, a day was up to 23 hours and over 30 minutes, and the year was down to 371 days. The moon would have looked a little larger and would have more strongly masked the sun during eclipses—there would have been none of the rare annular eclipses in which the moon is far enough away in its elliptical orbit that the sun rings the moon at maximum. The "man on the moon" leered down upon the dinosaur planet, but the prominent Tycho crater was not blasted into existence until toward the end

of the Early Cretaceous. As the sun converts an increasing portion of its core from hydrogen into denser helium, it becomes hotter by nearly 10 percent every billion years, so the sun was about 2 percent cooler when herbivorous dinosaurs first showed up and around 0.5 percent cooler than it is now when they went extinct.

At the beginning of the great Paleozoic era over half a billion years ago, the Cambrian Revolution saw the advent of complex, often hard-shelled organisms. Also appearing were the first simple vertebrates. As the Paleozoic progressed, first plants and then animals, including tetrapod vertebrates, began to invade the land, which saw a brief Age of Amphibians in the late Mississippian followed by the classic Age of Reptiles in the Pennsylvanian and much of the Permian. By the last period in the Paleozoic, the Permian, the continents had joined together into the C-shaped supercontinent Pangaea, which straddled the equator and stretched nearly to the poles north and south. Lying like a marine wedge between the terrestrial arms of northern Laurasia and southern Gondwanaland was the great Tethys tropical ocean; all that is left of that is the modest Mediterranean. Seventy percent of the world was dominated by the enormous Panthalassic superocean, whose vast expanse was marred only by occasional islands, some barely breaching the waves, others presumably large; the Pacific is its somewhat lesser descendant. With the majority of land far from the oceans, most continental habitats were harshly semiarid, ranging from extra hot in the tropics to sometimes glacial at high latitudes. The major vertebrate groups had evolved by that time. Among synapsids, the mammal-like therapsids, some up to the size of rhinos, were the dominant large land animals in the Age of Therapsids of the Late Permian. These were apparently more energetic than reptiles, and those living in cold climates may have used fur to conserve heat. Toward the end of the period, the first archosaurs appeared. These low-slung, vaguely lizard- or crocodilian-looking creatures were a minor part of the global fauna. The conclusion of the Permian was marked by a great extinction that is attributed to the massive volcanism then forming the enormous Siberian Traps—there is no evidence of a major meteorite impact at the time—which so heavily contaminated the atmosphere in multiple ways that the global environment terrestrial and marine was severely disrupted chemically and weatherwise. In some regards, the Permian–Triassic extinction exceeded that which killed off the land-bound dinosaurs 185 million years later.

At the beginning of the first period of the Mesozoic, the Triassic, the global fauna was severely denuded. As it recovered, the few remaining therapsids enjoyed a second evolutionary radiation and again became an important part of the wildlife. And again, they never became truly enormous or tall, although a few late examples were as big as small elephants. This time, they had serious competition, as the archosaurs also underwent an evolutionary explosion, first expressed as a wide variety of thecodonts, some of which reached a tonne in mass. One group evolved into aquatic, armored crocodile mimics. Others became armored land herbivores. Most were land predators that moved on erect legs achieved in a manner different from dinosaurs. The head of the femur did not turn inward; instead, the hip socket expanded over the femoral head until the shaft could be directed downward. Some of these erect-legged archosaurs were nearly bipedal. Others became toothless plant eaters. It is being realized that in many respects the Triassic thecodonts filled the lifestyle roles that would later be occupied by dinosaurs. Even so, these basal archosaurs never became gigantic or very tall. Also coming onto the scene were the crocodilians, the only group surviving today that reminds us what the archosaurs of the Triassic were like. Triassic crocodilians started out as small, long-legged, digitigrade land runners. Crocodilians, like many of the thecodonts, had a very undinosaurian feature: their ankles were complex, door-hinge-like joints in which a tuber projecting from one of the ankle bones helped increase the leverage of the muscles on the foot, rather as in mammals. In the Norian, a modest extinction event cut back somewhat on the diversity of therapsids and thecodonts. In the Late Triassic, the membrane-winged, long-tailed pterosaurs show up in the fossil record fully formed, since their evolution had started earlier. That pterosaurs had the same kind of simple-hinge ankle seen in dinosaurs is evidence that the two groups are related, forming the Ornithodira.

By the middle Triassic, quite small predatory archosaurs were evolving that exhibited a number of features of dinosaurs. Although the hip socket was still not internally open, the femoral head was turned inward, allowing the legs to operate in a predominantly vertical plane. The ankle was the simple fore-aft hinge that it remains in birds. The skull was lightly constructed. At first recorded as skeletal remains in the early Late Triassic of South America—trackways suggest they appeared earlier in the period—protodinosaurs have since been found on other continents. These bipedal/quadrupedal early near dinosaurs would survive only until the Norian, as they were displaced by their descendants. Protodinosaurs show that dinosaurs started out as little creatures; big dinosaurs did not descend from the big basal archosaurs.

From small things big things can evolve, and very quickly. In the Carnian stage of the Late Triassic, the fairly large-bodied, small-hipped, four-toed herrerasaur predators were on the global stage. These fully bipedal early dinosaurs with cylindrical hip joints dwelled in a world still dominated by complex-ankled archosaurs and would not last beyond the Norian or maybe the Rhaetian stage, perhaps because these early dinosaurs did not have an aerobic capacity high enough to vie with their newer dinosaurian competitors. The Norian saw the appearance of the great group that is still with us, the bird-footed avepod theropods, whose enlarging hips and beginnings of the avian-type respiratory system imply a further improvement in aerobic performance and thermoregulation—their story is detailed in a field guide on that group. At about the same time, the first members of the grand groups of herbivorous dinosaurs this guide is about are recorded in the fossil record: the small-hipped, semibipedal prosauropods with lightly built heads and leaf-shaped teeth best suited for tender vegetation, such as ferns. Whether prosauropods also had bird-style respiration is a matter of debate; if so, it was probably absent at first, and then weakly developed. Looking not all that different from their theropod contemporaries, the earliest prosauropods had short necks and lightly built arms and were in peril from their similarly sized predatory relation—even the small predaceous dinosaurs, whether juveniles or adults, would always menace juvenile sauropodomorphs. All these new dinosaurs gave therapsids and thecodonts increasing competition as they rapidly expanded in diversity as well as size. The prosauropods soon upgraded with the euprosauropods, which tended to be bigger and had the long necks needed to reach higher into the vegetation, maximizing upward reach by standing erect, their tail able to act as a third prop. Just 15 or 20 million years after the evolution of the first little protodinosaurs, rhino-sized prosauropods weighing 2 tonnes or more had developed. These long-necked dinosaurs were also the first herbivores able to browse at high levels, 5 m (15 ft) above the ground when rearing. Dinosaurs were showing the ability to evolve large dimensions and bulk on land, an attribute otherwise seen only among a few therapsids, some of which reached small elephant size in the Late Triassic but remained low-slung, low-level browsers. In the Carnian, the first of the beaked herbivorous ornithischians arrived. These little semibipeds were not common, and they, as well as small theropods and prosauropods, may have dug burrows as refuges from a predator-filled world. By the last stage of the Triassic, avepod and prosauropod dinosaurs were becoming the ascendant land animals, although they still lived among a number of thecodonts and some therapsids. From the latter, at this time, evolved the first mammals. Mammals and dinosaurs have therefore shared the planet for over 200 million years—but for 140 million of those years, mammals remained small.

Because animals could wander over the entire supercontinent with little hindrance from big bodies of salt water, faunas tended to exhibit little difference from one region to another. And with the continents still collected together, the climatic conditions over most of the supercontinent remained harsh. It was the greenhouse world that would prevail through the Mesozoic. The carbon dioxide level was 2–10 times higher than it is currently, boosting temperatures to such highs—despite the slightly cooler sun of those times—that even the polar regions were relatively warm in winter. The low level of tectonic activity meant there were few tall mountain ranges to capture rain or interior seaways to provide moisture. Hence, there were great deserts, and most of the vegetated lands were seasonally semiarid, but forests were located in the few regions of heavy rainfall and groundwater created by climatic zones and rising uplands. According to some research, the tropical latitudes were so hot and dry that the larger dinosaurs, with their high energy budgets, were hard-pressed to dwell near the equator and, except for tenuous coastal equatorial strips, were restricted mainly to the cooler, wetter, higher latitudes. The flora that early sauropodomorphs were dining on was in many respects fairly modern and included many plants with which we would be familiar. Wet areas along watercourses were the domain of lycophytes and horsetails. Some ferns also favored wet areas and shaded forest floors. Other ferns grew in open areas that were dry most of the year, flourishing during the brief rainy season. Large parts of the world may have been covered by fern prairies, comparable to the grasslands and shrublands of today. Tree ferns were common in wetter areas. Perhaps more abundant were the fernlike or palmlike cycadeoids, similar to the cycads that still inhabit the tropics. Taller trees included water-loving ginkgoaceans, of which the maidenhair tree is the sole—and, until widely planted in urban areas, the nearly extinct—survivor. Dominant among plants were conifers, most of which at that time had broad leaves, rather than needles. Some of the conifers were giants rivaling the colossal trees of today, such as those that formed the famed Petrified Forest of Arizona—these were beyond the reach of the not-yet-titanic prosauropods. Flowering plants were completely absent.

The end of the Triassic about 200 million years ago saw another extinction event. A giant impact had occurred in the hard, ancient rocks of southeastern Canada, but it was a few million years before the specific time of the faunal crisis. Coincident with the extinction were intense volcanic events tied to the formation of the budding Atlantic Ocean—note that volcanism was not particularly common in the Mesozoic relative to the subsequent Cenozoic.

The Late Triassic *Buriolestes* and *Staurikosaurus*

The heavy dusting of the atmosphere with reflective aerosol debris cooled off the planet to the degree that minor glaciations occurred at high latitudes. The thecodonts and therapsids suffered the most: the former were wiped out with the exception of crocodilians, and only scarce remnants of the latter survived along with furry mammal relatives. The more energetic, sometimes-insulated pterosaurs and dinosaurs were able to deal with the big chills without difficulty and sailed through the crisis into the Early Jurassic with little disruption, so much so that avepod theropods and euprosauropods remained common and little changed, although the short-necked basal prosauropods did not make into in the new period. For the rest of the Mesozoic, dinosaurs, including those herbivorous, would enjoy almost total dominance on land except for some substantial semiterrestrial crocodilians; otherwise there simply were no competitors above a few kilograms in weight. Such extreme superiority was unique in Earth's history. The Jurassic and Cretaceous combined were the Age of Dinosaurs.

The literally big novelty of the Early Jurassic was the advent of the sauropods that had descended from euprosauropods—recent claims that sauropods first appeared in the Late Triassic have not been borne out. Some of the quadrupeds that were elephantine in both size and body and leg form, with very long necks and tails added, got as big as bull African elephants early in the Jurassic. Never truly small-bodied, their columnar legs meant these were the first dinosaurs unable to achieve a full run, so being big was necessary for defense, among other reasons. At first, the euprosauropods did well. They were even in snowy Antarctica. Some remained little different from their Triassic predecessors; others became heavier-armed and shorter-skulled. The biggest known examples were as bulky as female Asian elephants at over 3 tonnes and able to reach as high as giraffes at over 6 m (20 ft). But as the Jurassic progressed, prosauropods appear to have been unable to compete with the more sophisticated, big-hipped sauropod they had spawned and were gone by the end of the Early Jurassic. Also out were the early avepod theropods in

The Early Jurassic *Anchisaurus* and *Podokesaurus*

favor of the averostrans with bigger hips and more power-ful leg muscles that began to show up in the Early Jurassic. It is likely that the loss of Triassic-grade dinosaurs in favor of more modern sauropods and averostrans was finalized by another bout of mass volcanism, this time in southern Gondwana, in the late Early Jurassic, which is associated with tough conifers becoming the dominant trees during the rest of the Jurassic into the Cretaceous. Sauropods' stouter skulls and teeth atop stronger necks apparently rendered them better able to deal with the harder-to-chew flora, as did their more capacious digestive tracts, which were better able to break down fodder. Larger hip mus-cles and an increasingly well-developed, birdlike respira-tory system suggest that sauropods had the higher aerobic capacity and higher-pressure circulatory system needed to achieve truly great height and tonnage. Although some theropods were getting moderately large, the much more gigantic adult sauropods—as heavy as bull African ele-phants and achieving a rearing height of 8 m (25 ft)—en-joyed a period of relative immunity from attack. This may have been aided by small clubs that adorned the tips of some sauropod tails. Ornithischians remained uncommon and small, albeit able to complete with smaller and juve-nile sauropodomorphs, and one group was the first set of dinosaurs to develop armor protection. On the continents, crocodilians remained small and fully or partly terrestrial, with some being herbivorous.

Farther west, the supercontinent was beginning to break up, creating African-style rift valleys along today's East-ern Seaboard of North America that presaged the opening of the Atlantic. Even so, Europe remained an Indonesia-like archipelago of islands immediately northeast of North America, as it had been for a long while. More importantly for dinosaur faunas, the increased tectonic activity in the continent-bearing conveyor belt formed by the mantle caused the ocean floors to lift up, spilling the oceans onto the continents in the form of shallow seaways that began to isolate different regions from one another, encouraging the evolution of a more diverse global wildlife. The expan-sion of such large water surfaces onto the continents also raised rainfall levels, although most habitats remained sea-sonally semiarid. The less extremely hot climes allowed big dinosaurs to roam about at all latitudes. The movement of the landmasses also produced more mountains able to squeeze rain out of the atmosphere. The opening of the At-lantic meant that the now-shrinking Panthalassic colossus was becoming the almost-as-great Pacific.

Beginning 175 million years ago, the Middle Jurassic began the long Age of Sauropods, whose increasingly so-phisticated respiratory and circulatory systems allowed them to match medium-sized whales in bulk and trees in height. Sauropods thrived even in dry habitats by feed-ing on the open, conifer-dominated forests that spread over flats and lined watercourses, as well as the fern prai-ries in the wet season. One sauropod group, the African atlasaurs, put emphasis on arm length over neck length to increase feeding height, but this experiment went no further, as elongating necks became the norm. In China, partly isolated by seaways, some mid-Jurassic sauropods evolved slender necks so long that they could feed 10 m high (over 30 ft), and by the end of the stage, the height maximum had topped off at six to seven stories on ani-mals of three dozen tonnes. A few examples had small tail clubs with which to pummel opponents. Thumb claws re-mained large weapons. Also appearing were the first small, armored stegosaur ornithischians that also introduced tail spikes. Although the increasingly sophisticated theropods evolved and featured highly developed avian-type respira-tory systems, for reasons that are obscure, they continued to fail to produce true giants that could take on big sau-ropods; those yet known reached barely a tonne at most. There is tenuous evidence that flowering plants were pres-ent by the middle of the Jurassic, but if so, they were not yet common.

The Late Jurassic, which began 160 million years ago, was the apogee of two herbivorous dinosaur groups: the sauropods and the stegosaurs. Sauropods—which in-cluded mamenchisaurs, turiasaurs, haplocanthosaurs, dicraeosaurs, diplodocines, apatosaurines, camarasaurs, brachiosaurs, and others—would never again be so di-verse. Some had short, deep heads with rows of big, spat-ulate teeth, others domed nasals, others low heads with pencil teeth restricted to the front end of the jaw. Necks and tails ranged from super long down to on the short side, the first sometimes ending in whips. Limbs fore and aft might be similar in length, resulting in level shoulders, or shoulders were low because of short arms and hands and big hips, or set high atop elegantly long forelimbs and hands. Thumb claws ranged from very big down to mod-estly sized in brachiosaurs, as the trend to their reduction began. Big-hipped, short-armed diplodocids were adept at both grazing ground cover and rearing on legs and tail to feed high. Camarasaur hips were tilted back to allow them to slow walk on two legs while they fed in tree crowns. Camarasaurs and brachiosaurs include the first broad-bel-lied sauropods, as the group shifted toward higher-capac-ity, fodder-fermenting digestive complexes. Many of these types tended to be widely distributed on the still closely connected continents, forming a global sauropod commu-nity. A given ecosystem of the great Morrison Formation—the geographically largest of all time—could include a half dozen or more taxa. The sauropods of the westernmost Europe closest to North America were very similar to, and in some cases perhaps the same as, those of the Morrison. At the same time, the sauropods of southern Africa were both similar to and distinctive from those up north. Some neosauropods rapidly enlarged to 50–75 tonnes, and a few may have greatly exceeded 100 tonnes, rivaling the biggest baleen whales. But it was a time of growing danger for the sauropods: theropods had finally evolved hippo-sized meg-alosaurs, yangchuanosaurs, and allosaurs that could tackle the colossal herbivores. Meanwhile, some sauropods isolated on islands underwent dwarfing to rhino size to accommodate to the limited resources (the same would later happen to island elephants and hippos). The now rhino- and sometimes elephant-sized stegosaurs were at their most diverse, with a few mimicking sauropods a little with their fairly long necks. And stegosaurs could rear up to increase browsing reach. Having evolved columnar legs, these dinosaurs also could not run. The other group of big, armored dinosaurs, the armadillo-like, short-legged ankylosaurs, was beginning to develop. Also entering the fauna were the first fairly large ornithopods. Appearing on the paleo scene were the first known theropods, the elaphrosaurs, which shifted away from dining on flesh toward plants, as competition with at least the lesser size zone of sauropods ramped up. At the small end of things, winged protobirds were beginning to appear. Pterosaurs remained fairly small-bodied. Although some crocodilians were still small runners, the highly amphibious crocodilians of the sort we are familiar with were appearing—but not particularly large. Although small, mammals were undergoing extensive evolution in the Jurassic. Dinosaurs did not yet show signs of being specifically adapted for life in the water, for reasons that remain obscure. There never would be dinosaurian hippos.

During the Middle and Late Jurassic, carbon dioxide levels were incredibly high, with the gas making up between 5 and 10 percent of the atmosphere, 10–20 times twenty-first-century levels (which are about twice preindustrial norms). As the Jurassic and the classic Age of Sauropods ended,

the incipient North Atlantic was about as large as today's Mediterranean. Vegetation had not yet changed dramatically from the Triassic. Wetter areas were dominated by conifers similar to cypress. A widespread and diverse conifer group of the time was the araucarians. Some appear to have evolved a classic umbrella shape in which most of the adult trunk was as bare of foliage as a telephone pole, with all the branches concentrated at the top. Still seen in some South American examples, this odd shape may have evolved as a means of escaping browsing by the ever-hungry sauropods, which should have had a profound impact on floral landscapes, as they heavily browsed and wrecked trees to an extent that probably exceeded that of elephants. What happened to the fauna toward the end of the Jurassic is not well understood because of a lack of deposits. Some researchers think there was a major extinction, in part because new data show that the very different Early Cretaceous dinofaunas were up and running closer to the Jurassic/Cretaceous boundary than was thought, maybe even a little before, indicating a sudden and fast turnover of the type that is usually evidence of a major global disruption. For example, why did the sophisticated diplodocids, with their exceptional feeding flexibility and self-defense capacities vis-à-vis thumb claws and whip tails, go extinct, especially when the titanosaurs would later also evolve horselike heads with pencil teeth, and whip tails? It is possible that the medium-capacity abdominal tracts of diplodocids were holding them back. Possible causal events include a giant crater dug by an impactor in South

The Late Jurassic *Dicraeosaurus* and *Giraffatitan*

Africa shortly before the end of the Jurassic, and mass vulcanism in the northwest Pacific.

Whatever did or did not happen, the epic Cretaceous began some 145 million years ago. This 80-million-year-long period would see an explosion of dinosaur evolution that surpassed all that had gone before, as the continents continued to split, the south Atlantic began to open, India detached from Antarctica and then Madagascar, as it began its epic voyage toward Asia, and seaways crisscrossed the continents. Greenhouse conditions became less extreme as carbon dioxide levels gradually edged downward, with a quick, sharp drop near the end of the Valanginian that looks like it led to polar glaciers on a limited scale. Not that carbon dioxide levels dropped to the modern preindustrial level. Early in the Cretaceous, the warm Arctic oceans generally kept conditions up there balmy even in winter. At the other pole, continental conditions rendered winters frigid enough to form permafrost. General global conditions were a little wetter than they were earlier in the Mesozoic, but seasonal aridity remained the rule in most places, and true rain forests continued to be scarce at best.

Sauropods remained abundant and often enormous, but they were less diverse than before, as a few small-bodied, short-necked rebbachisaurid diplodocoids—some with broad, square-ended mouths specialized for grazing—tall brachiosaurs, and especially the broad-bellied titanosaurs predominated. Yet their time of overwhelming dominance was past to the degree that the Cretaceous was the Age of Ornithischians. Ornithopods small and especially large flourished. Among the larger examples, well-developed plant-pulping dental batteries that paralled those of many mammals of the Cenozoic, while being entirely different from the sauropod system, may have been a key to the ornithopods' success. Iguanodont ornithopods soon became common large herbivores in the Northern Hemisphere. Stegosaurs soon departed the planet, the second and final major herbivorous dinosaur group to become extinct after the prosauropods. In the place of stegosaurs, the low-slung and extremely fat-bellied, armored ankylosaurs became a major portion of the global fauna, their plates and spikes providing protection from the big Laurasian allosauroids that were capable of taking on sauropods, and snub-nosed, short-armed abelisaurs in Gondwana. Among theropods, two new groups went herbivorous: the toothless ornithomimosaurs, most of which were ostrich-like in form and size, and the enigmatic, potbellied, plant-chewing therizinosaur dinosaurs, whose heads and fairly long necks were reminiscent of prosauropods. So were their big finger and two claws with which to lash out at enemies. Among the lesser theropods, the powerfully armed, sickle-clawed, bird-like dromaeosaurs were able to attack fair-sized, growing sauropods. Full-blown birds were appearing, no doubt occasionally riding on the backs of sauropods.

Pterosaurs were becoming large, as they met increasing competition from birds. Also fast increasing in size were the freshwater crocodilians, making them an increasing threat for young sauropods coming to water to drink or for other purposes. Some large crocodilians were semiterrestrial and able to attack big dinosaurs on land as well as in the water. Still scampering about were a few small

The Early Cretaceous *Nigersaurus* and *Ouranosaurus*

running crocodilians. Some carnivorous mammals were big enough, about a dozen kilograms (25 lb), to put baby sauropods at risk.

During the late Early Cretaceous, a major evolutionary event occurred, one that probably encouraged the rapid evolution of herbivorous dinosaurs: flowering plants began to become an important portion of the global flora. The first examples were small shrubs growing along shifting watercourses, where the plants' ability to rapidly colonize new territory was an advantage. Others were more fully aquatic, including water lilies. Their flowers were small and simple. The fast growth and strong recovery potential of flowering plants may have encouraged the development of low-browsing ankylosaurs and ornithopods. Conversely, the browsing pressure of dinosaurs may have been a driving force behind the evolution of the fast-spreading and fast-growing new plants. In the middle of the Cretaceous, ground-covering grasses appear to have been coming onto the scene, perhaps in regions below the equator, starting to evolve the hard silica phytoliths that hinder grazing by abrading teeth.

In the Late Cretaceous, which began 100 million years ago with the continental breakup well under way, interior seaways often covered vast tracts of land. Early on, temperatures picked up to high levels, but as carbon dioxide levels again dropped, the dark Arctic winters became cold enough to match the conditions seen in today's high northern forests, and glaciers crept down high-latitude mountains. Early in the Late Cretaceous, the moisture-loving dawn redwood, a now rare and atypically deciduous conifer, evolved to cope with the dark circumpolar winters. Mammals were increasingly modern—and still small. Pterosaurs became gigantic to a degree that stretches credulity. The continental azhdarchids sported wings of 11 m (over 35 ft) and easily outweighed ostriches; they could have picked up little titanosaurs that had recently departed their nests. Small running crocodilians remained extant, some herbivorous. As for the conventional freshwater crocodilians, in some locales they became colossi up to 12 m long (close to 40 ft) and approaching 10 tonnes, as large as the biggest flesh-eating theropods. Although these monsters fed mainly on fish and smaller tetrapods, they posed a real threat to all but the largest dinosaurs when they came to large bodies of water to drink or swam across large bodies of water. The hazard should not be exaggerated, however, because these super-crocs do not appear to have been very numerous in many locations, were absent at higher latitudes, and did not last all that long. Even so, their existence may have discouraged the evolution of profoundly aquatic dinosaurs, none of which went fully aquatic, much less marine.

As the Late Cretaceous began, a few rebbachisaurids were still on hand, some fairly large. But sauropods soon became limited to broad-beamed bellies, as titanosaurs alone proliferated across most of the globe, being especially diverse in the Southern Hemisphere, wrapping up the 150 million years that made the clade the most successful herbivore group in Earth's history. Titanosaurs were themselves fairly diverse. A few had short, deep heads. But most sported more-appealing-looking, long, low, horselike crania with slender teeth toward the front of the jaw remarkably like those of diplodocoids in a classic example of evolutionary convergence. Necks were from medium to quite long, but not exceptionally so. Arms varied from about the same length as the legs to somewhat longer, pitching the shoulders up some in the latter cases. Thumb claws became lost entirely. Weaponry was correspondingly limited to tails. Ball-and-socket joints at the base made titanosaur tails extra flexible, and the caudals ended in short whips. Titanosaurs disappeared from North America for part of the Late Cretaceous, only to reappear in the drier regions toward the end. Why this sauropod hiatus occurred remains mysterious; there had been plenty of plants on the continent for the beasts to gulp down. Appalachia, the isolated eastern portion of North America, was a large territory. Some titanosaurs were lightly armored; this may have in part been a means to protect the juveniles against the increasing threat posed by a growing assortment of predators. A few rather small titanosaurs had the short necks and square, broad mouths suited for grazing. Those from European islands look like dwarfs. Others were titanic, exceeding 50 and reaching perhaps 200 tonnes—as big as the biggest whales—up to the end of the dinosaur era. That indicates that some sauropods were as massive as, and much longer than, the greatest titans of the seas for 90 million years. Titanosaurs, supersized and otherwise, were subject to attack from abelisaur and allosauroid theropods, some of the latter matching bull elephants in bulk—that until the allosauroids went out of existence in the early Late Cretaceous.

Ornithomimosaurs included the fleet-footed ornithomimids and the many-tonne deinocheirids of late Late Cretaceous Asia. Some of the erect-bodied therizinosaurs also got to 5 tonnes, maybe 10, making them the most direct competitive peers of titanosaurs. The ultrawide-bodied ankylosaurs continued their success, especially in the Northern Hemisphere. Among big ornithopods, the iguanodonts faded from the scene to be replaced by their fleeter descendants, the duck-billed hadrosaurs, which evolved the most complex grinding dental batteries among dinosaurs, if not all herbivores, and had elaborate head crests that identified the variety of species. The most common herbivores in much of the Northern Hemisphere, hadrosaurs may have been adapted in part to browse on the flowering shrubs and ground cover that were beginning to replace the fern prairies as well as to invade forest floors. A few hadrosaurs of Asia and North America posed the only serious size competition to sauropods, approaching 15 tonnes. The rhino- to elephant-sized ceratopsids bore oversized heads with great parrotlike beaks and slicing dental batteries. These ornithischians flourished for just the last 15 million years of the dinosaur era, limited largely to the modest-sized stretch of North America that lay west of the interior seaway; their presence in Asia was very restricted.

The Late Cretaceous *Utetitan*, *Tyrannosaurus*, and *Quetzalcoatlus*

Birds, some still toothed, others representing the beginnings of modern avians, continued to thrive. Dromaeosaurs able to take down growing titanosaurs remained in force. Culminating more than 150 million years of big avepod history were the great tyrannosaurids, the most sophisticated and powerful of the gigantic sauropod killers. Tyrannosaurids came into existence only some 15 million years before the end of the Mesozoic and were limited to Asia and North America—on the latter continent, they were a problem only for the titanosaurs of the Southwest. Those have long been lumped into a single taxon despite being extant for a few million years, but the latest work shows that they were more diverse than that, representing a radiation of their own. The factors that limited North American titanosaurs to that region are not clear. No physical barriers kept them from moving north into lands that were still subtropical, but they did face off with southerly populations of mighty *Tyrannosaurus*. Appalachia was out of reach until the withdrawing interior seaway opened it up to colonization shortly before the Cretaceous ended; there is no current evidence sauropods made it there. It is interesting that the scrappy titanosaur remains from the latest Mesozoic, which may represent the largest land animals known to date, are from India, which was a big island at the time. By the end of the Cretaceous, the continents had moved far enough that the world was assuming its modern configuration. At the terminus of the period, a burst of uplift and mountain building had helped drain many of the seaways, although Europe remained an island archipelago. Flowering plants were fast becoming an ever more important part of the flora, and the first hardwood trees—among them the plane tree commonly planted in cities—appeared near the end of the period and were evolving into the first large hardwood trees. Also gracing landscapes were true palms. Conifers remained dominant, however, including the classic moisture-loving, fast-spreading, rapidly growing redwoods, which reached towering heights, as well as sequoias, as they do today. In South America, the browsing burden of the towering titanosaurs may have continued to encourage the evolution of the umbrella-topped monkey-puzzle araucarians. Classic rain forests, however, still did not exist. Titanosaur dung shows they were eating pytolith-containing grasses that were beginning to spread out onto dry flats at the long-term expense of ferns and herbaceous ground cover.

Then things went catastrophically wrong.

EXTINCTION

The mass extinction at the end of the Mesozoic is generally seen as the second most extensive in Earth's history, after the one that ended the Paleozoic. However, the earlier extinction did not entirely exterminate the major groups of large land animals. At the end of the Cretaceous, all nonavian dinosaurs, the only major land animals, were lost, leaving only flying birds as survivors of the group. Among the birds, all the toothed forms, plus the major Mesozoic bird branch, the enantiornithines, as well as the flightless birds of the time, were also destroyed. So were the last of the superpterosaurs and the most gigantic of the crocodilians.

It is difficult to exaggerate how remarkable the loss of the dinosaurs was. If they had repeatedly suffered the elimination of major groups and experienced occasional diversity squeezes, in which the Sauropoda and other Dinosauria were reduced to a much smaller collection that then underwent another evolutionary radiation until the next squeeze, then their final loss would not be so surprising. But the opposite is the case. A group that had thrived for over 150 million years over the entire globe, rarely suffering the destruction of a major subgroup and usually building up diversity in form and species over time as it evolved increasing sophistication, was in very short order completely expunged. The small dinosaurs went with the large ones, predators along with herbivores and omnivores, and intelligent ones along with those with reptilian brains. It is especially notable that even the gigantic dinosaurs did not suffer repeated extinction events. Sauropods were always a diverse and vital group for most of the reign of dinosaurs. Only the prosauropods—probably outcompeted by early sauropods—and stegosaurs had faded away well into the dinosaur era. In contrast, many of the groups of titanic Cenozoic herbivorous mammals (arsinoitheres, uintatheres, brontotheres, paraceratheres) appeared, flourished relatively briefly, and then went extinct. Dinosaurs appear to have been highly resistant to large-scale extinction. A reason for that may be the way they reproduced. Whether small or great, the sauropods included, dinosaurs were fast breeders that laid large numbers of eggs, making them weed species that enjoyed a high potential for population recovery and expansion as long as at least some of the many eggs and/or juveniles survived a severe crisis and could continue the species. In contrast, big mammals reproduce slowly, dropping a calf only once a year or less, and that offspring then needs the careful attention of a nursing parent to grow up. That long parenting time investment leaves slow-breeding species highly vulnerable to elimination if the ecosystem temporarily becomes toxic. Rendering their elimination still more remarkable is that one group of dinosaurs, the birds, did survive, as well as aquatic crocodilians, lizards, snakes—the latter had evolved by the Late Cretaceous—amphibians, and mammals that proved able to weather the same crisis.

It has been argued that dinosaurs were showing signs of being in trouble in the last few million years before the final extinction. Whether they were in decline has been difficult to verify or refute even in those few locations where the last stage of the dinosaur era was recorded in the geological record, such as western North America, where the fossil data are the best. Even if these declines did occur, they were modest, and other regions of the globe may not have seen a decrease. At the Cretaceous/Paleocene (K/Pg) boundary—formerly the Cretaceous/Tertiary (K/T) boundary—the total population of juvenile and adult dinosaurs should have roughly matched that of similarly sized land mammals before the advent of humans, numbering in the billions and spread among many dozens or a few hundred species on all continents and many islands. Titanosaurs would have been a substantial minority of that population.

A changing climate has often been offered as the cause of the dinosaurs' demise. But the climatic shifts at the end of the Cretaceous were neither strong nor greater than those already seen in the Mesozoic. If anything, reptiles should have been more affected. The rise of the flowering plants has been suggested to have harmed dinosaurs, but the increase in food provided by the fast-growing seed- and fruit-producing plants appears to have been so beneficial that it spurred the evolution of late Mesozoic dinosaurs, although it may not have had much effect on the titanosaurs. Mammals consuming dinosaur eggs were another proposed agent. But sauropods had been losing eggs to mammals, and for that matter lizards and the like, for nearly 135 million years, and so had reptiles and other dinosaurs, including birds, without long-term ill effects. The spread of diseases as retreating seaways allowed once-isolated dinosaur faunas to intermix is not a sufficient explanation because of the prior failure of disease to crash the sauropod population, which was too diverse to be destroyed by one or a few diseases and which would have developed resistance and recovered its numbers. Also unexplained is why other animals survived.

The solar system is a shooting gallery full of large rogue asteroids and comets that can create immense destruction. There is widespread agreement that the K/Pg extinction was caused largely or entirely by the impact of at least one meteorite, a mountain-sized object that fell from the northeast at a steep angle and formed a crater 180 km across (over 100 mi), located on the Yucatán Peninsula of Mexico. The evidence strongly supports the object being an asteroid, rather than a comet, so speculations that a perturbation of the Oort cloud as the solar system traveled through the galaxy and its dark matter are at best problematic. The explosion of 100 teratons surpassed the power of the largest H-bomb detonation by a factor of 20 million and dwarfed the total firepower of the combined nuclear arsenals at the height of the Cold War. The blast and heat generated by the explosion wiped out the fauna in the surrounding vicinity, and enormous tsunamis, either arriving directly from the impact area or locally induced by the massive worldwide earthquakes emanating from the massive hit, cleared off coastlines up what was left of the interior seaway, across the Atlantic and even the Pacific. The cloud of high-velocity debris ejected at near-orbital velocities into near space glowed hot as it reentered the atmosphere in the hours after the impact, creating a world-wrapping pyrosphere that may have been searing enough to bake animals to death as it ignited planetary wildfires. A fossil site exactly at the K/Pg divide in North Dakota appears to record some of these events, including remains of dinosaurs possibly drowned by the local flooding and impact debris in the gills of fish. The latter suggests it was a springtime disaster. This location does not involve titanosaurs, since they were missing from that region's fauna. The initial impact would have been followed by a solid dust pall that plunged the entire world into a dark, cold winter lasting for years, combined with severe air pollution and acid rain. As the aerial particulates settled, the climate then flipped, as enormous amounts of carbon dioxide created what some researchers reconstruct as an extreme greenhouse effect that baked the planet for many thousands of years. (The gas was released because the impact happened to hit a tropical marine carbonate platform and dug deeply into it at a steep angle—a more glancing impact or one elsewhere might not have had such serious effects.) Such a combination of agents appears to solve the mystery of the annihilation of the titanosaurs et al. The modern-style birds that got through, albeit barely, may have done so because they brooded their eggs, protecting them from the dreadful climate, and had flexible kinetic bills that aided feeding on whatever was left to eat. Even so, this scenario remains somewhat problematic.

As big as the asteroid was, it was the size of a mere mountain and was dwarfed by the planet it ran into. It is not certain whether the resulting pyrosphere was as planetarily lethal as some estimate. Even if it was, heavy storms covering a small percentage of the land surface should have shielded a few million square kilometers, in total equaling the size of India, creating numerous scattered refugia. In exposed locations, dinosaurs, including juvenile titanosaurs, that happened to be in shallow waters not subject to big wave action should have survived the pyrosphere like the lucky crocodilians and amphibians. So should many of the eggs buried in covered nests. Birds and amphibians, which are highly sensitive to environmental toxins, survived the acid rain and pollution. Because titanosaurs were rapidly

reproducing "weed species" whose self-feeding young could survive without the care of parents, at least some dinosaur populations should have made it through the crisis, as did some other animals, rapidly recolonizing the planet as it recovered. Recent work challenges the extremity of the asteroid winter and the following hothouse.

A major complicating factor is that massive volcanism occurred at the end of the Cretaceous, as enormous lava flows covered 1.5 million square kilometers, a third of the Indian subcontinent. Had this not been going on, the extinction could be blamed on the impact, whereas if the latter had not occurred, then the volcanism could be charged. Nor is it likely that bouts of intense volcanic events are a regular occurrence. As it is, the extinction waters are muddied—similar to how the atomic bombings of Japan and the massive Soviet attack at the same time mean that the surrender cannot simply be assigned to the former. It has been proposed that the massive air pollution produced from the repeated supereruptions of the Deccan Traps damaged the global ecosystem so severely in so many ways that dinosaur populations collapsed in a series of stages, perhaps spanning tens or hundreds of thousands of years. This hypothesis is intriguing because extreme volcanic activity was responsible for the great Permian–Triassic extinction, and other of the infrequent mass eruptions appear to have caused serious losses. Although the K/Pg Deccan Traps were being extruded before the Yucatán impact and in the process were possibly degrading the global fauna and flora, evidence indicates that the impact—which generated earthquakes of magnitude 9 over most of the globe (11 on the exponential scale at the impact site)—may have greatly accelerated the frequency and scale of the eruptions. If this is correct, then the impact was responsible for the extinction not just via its immediate, short-term effects, but by sparking a level of extra intense supervolcanism that prevented the recovery of the dinosaurs. It is also possible that the Yucatán impactor was part of an asteroid set that hit the planet repeatedly, further damaging the biosphere.

Even the combined impact/volcanic hypothesis does not fully explain why dinosaurs failed to survive problems that other continental animals did. But being so large as adults, titanosaurs were not in the most optimal position to survive the ecological crisis.

Although extraterrestrial impact(s), perhaps indirectly linked with volcanism, is the leading explanation, the environmental mechanisms that destroyed all the nonflying dinosaurs while leaving many birds and other animals behind remain incompletely understood.

AFTER THE TIME OF TITANOSAURS

Perhaps because trees were freed from chronic assault by multistory titanosaur sauropods, dense forests, including rain forests, finally appeared. In the immediate wake of the extinction, there were no large land animals, and only

large freshwater crocodilians could make a living feeding on fish. The loss of dinosaurs led to a second, brief Age of Reptiles, as superboa snakes as long as the biggest theropods and weighing over a tonne quickly evolved in the

Giant *Paraceratherium*

tropics, which also sported big freshwater turtles. By 40 million years ago, about 25 million years after the termination of large dinosaurs, some land mammals were evolving into giants rivaling the latter. But only to a degree. No mammals rivaled sauropods in becoming extra tall tree browsers able to feed many stories in height; long-necked giraffes and long-trunked elephants can reach only a couple of stories. Where mammals have matched dinosaurs is in the oceans; there whales have reached and exceeded 100 tonnes. But while sauropods did that for 90 million years, supersized whales actually appeared only a few million years ago in association with the ongoing ice age.

BIOLOGY

GENERAL ANATOMY AND LOCOMOTION

Sauropodomorphs come in the two basic and distinctive types: prosauropods and sauropods. The first were fairly uniform in form; sauropods differed tremendously from one another in body plan. All were plant eaters, although they were not averse to consuming some animal protein. Heads were small, but this needs to be qualified. Crania were always small relative to the rest of the animal and were absolutely so in prosauropods and smaller sauropods, whether adult or juvenile. But the heads of the greater sauropods were attached to huge bodies, and the comparatively small heads were quite hefty in actual dimensions.

Sauropod skull and muscles

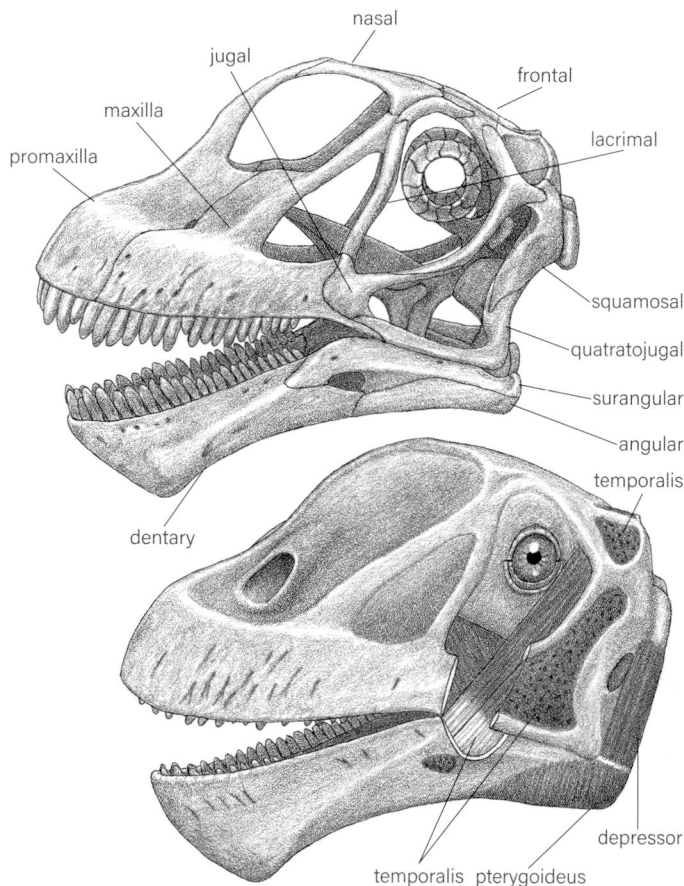

One researcher claimed that the heads of 40 tonne brachiosaurs were similar in size to those of 1 tonne giraffes, but the mouth of the sauropod could engulf the head of a giraffe and could swallow a child whole—the kids in the treetops in *Jurassic Park* were in more danger from the brachiosaurs than they realized. The heads of the larger sauropods were a meter more in length and half a meter plus in depth.

Heads ranged from delicately constructed, especially in some prosauropods, to fairly solidly built. In all examples, the nasal passages and sinuses were very well developed, a feature common to archosaurs in general. A distinctive feature of prosauropods is that the nasals are somewhat inflated laterally, as can be seen in top view. All prosauropods and sauropods retained a large opening immediately in front of the orbits, this generally being larger in the first. Unlike mammals, with their extensive facial musculature, dinosaurs, like reptiles and birds, lacked extensive facial muscles, so the skin was directly appressed to the skull. This feature makes dinosaur heads easier to restore than those of mammals. The external nares are always located far forward in the nasal depression, no matter how far back on the skull the nasal openings extend. In some sauropods, the nasal openings are set far back on the skull, above the eye sockets. It was once thought that this allowed these dinosaurs to snorkel when submerged. More recently, it has been suggested that the retracted nostrils evolved to avoid irritation from needles as sauropods fed on conifers. Most conifers at that time, however, had soft leaves. In any case, the fleshy nostrils probably extended far forward so that the openings were in the normal position near the tip of the snout—there is no anatomical evidence that any sauropodomorph had a proboscis, as had been suggested for some sauropods. The skin covering the large openings in front of the orbits of many dinosaurs would have bulged gently outward, since they were filled with sinuses. While jaw muscles likewise bulged subtly out of the skull openings aft of the eye sockets, these were not shallow depressions, as some artists show.

Head depth ranged from shallow to deep. The last was true of diplodocoids and titanosaurs, in which the joint for the lower jaw was set well forward, below the orbits, resulting in short jaws. Those heads had a distinctive, horselike look. In top view, sauropodomorph heads varied from fairly narrow to fairly broad. In the latter, the jaw muscles were powerful enough to crop tough vegetation. Prosauropod mouths were fairly sharp-tipped in top and bottom view. Sauropod mouths were blunter, usually being rounded. In

Sauropod skeletal, muscle, and life restoration

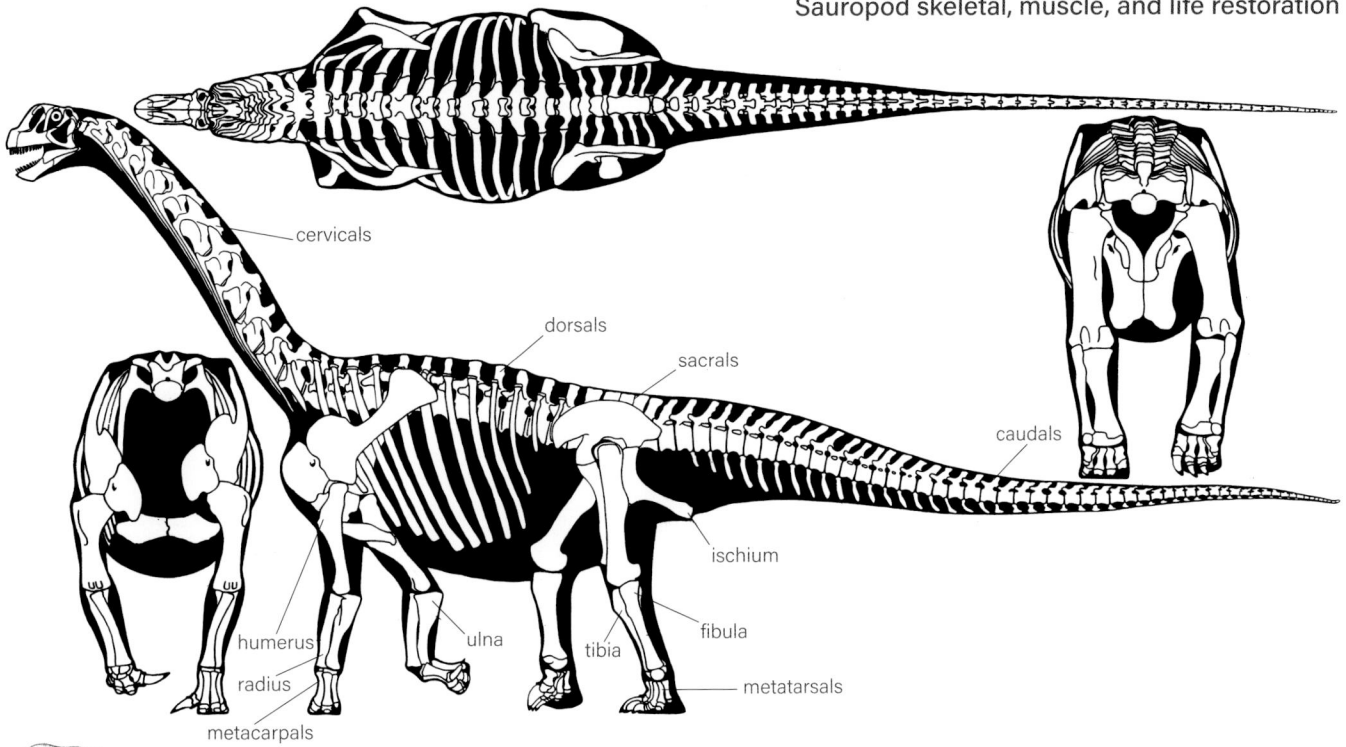

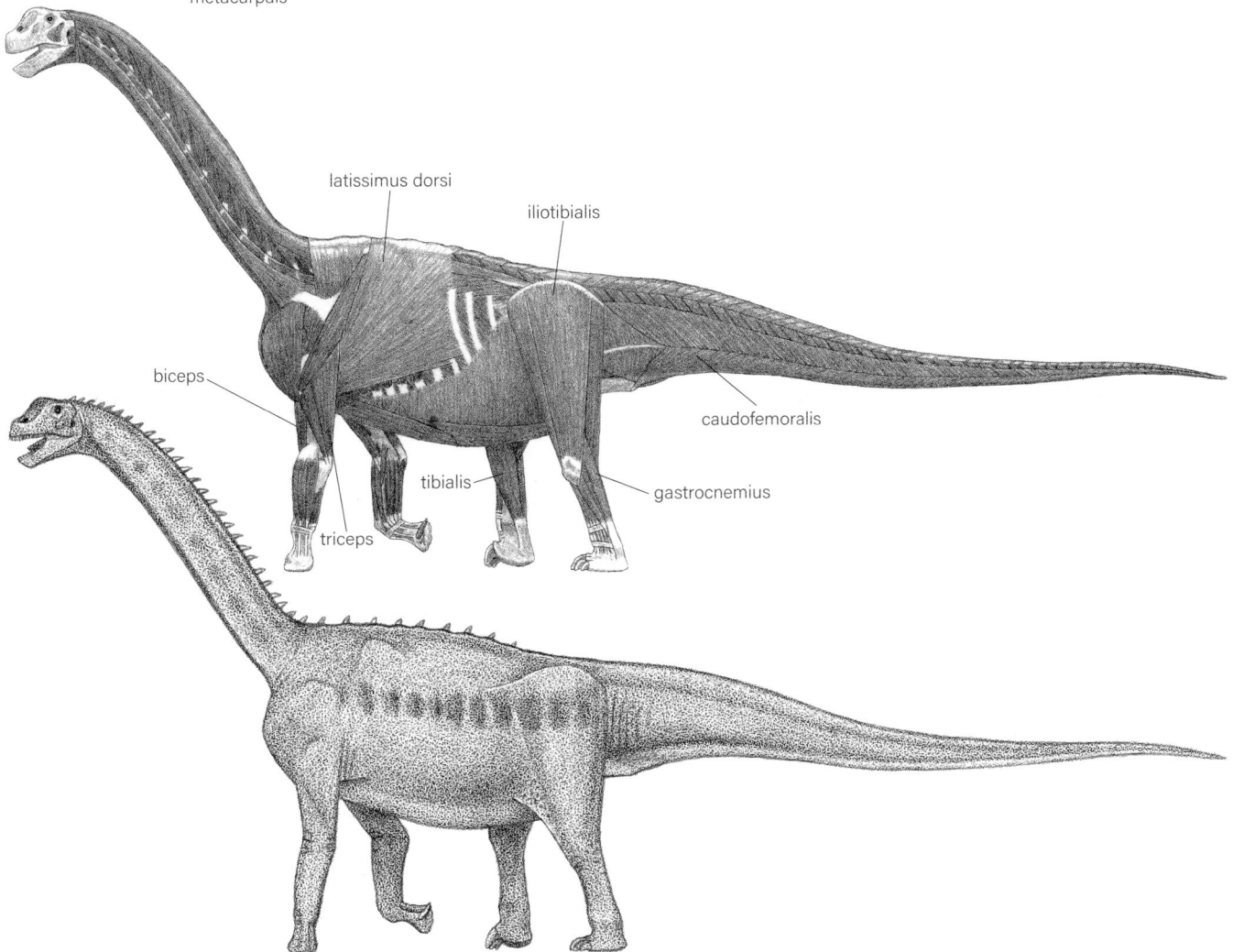

cervicals

dorsals

sacrals

caudals

ischium

humerus

ulna

tibia

fibula

radius

metatarsals

metacarpals

latissimus dorsi

iliotibialis

biceps

caudofemoralis

tibialis

gastrocnemius

triceps

Teeth, actual size

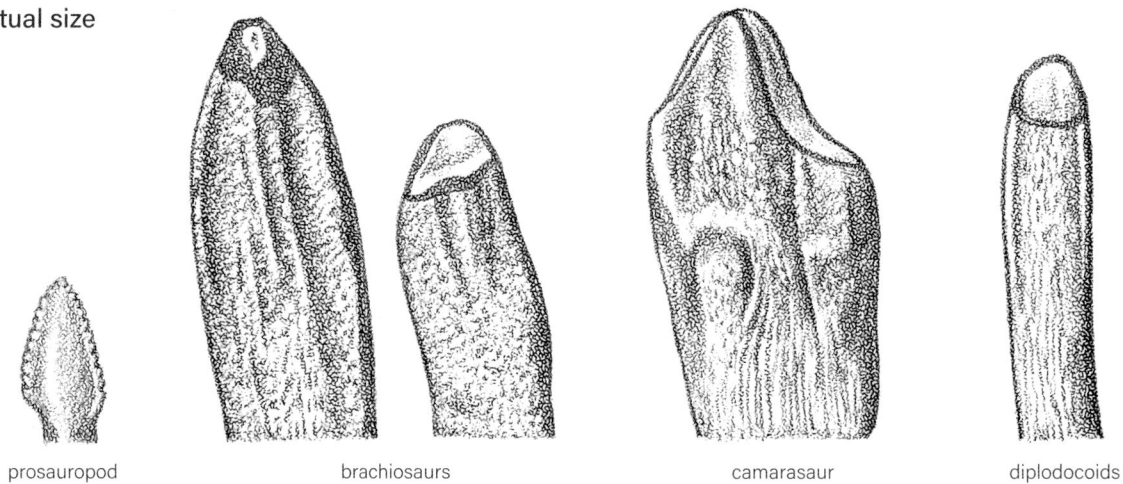

prosauropod   brachiosaurs   camarasaur   diplodocoids

no sauropodomorphs did the teeth form tightly occluding rows like those of large ornithopods and ceratopsids or ungulate mammals. The earliest, short-necked prosauropods sometimes featured a mixture of sharp and blunt teeth, reflecting an omnivorous diet that was transitional between the flesh eating of their basal dinosaur ancestors and the herbivory the group was evolving toward. Among all euprosauropods and many sauropods, the teeth were leaf-shaped, with blunt tips suited for ripping and pulping plant materials. In some sauropods, the teeth were fairly large. Tooth crowns often bore large, rounded serrations. The teeth formed long rows lining much of the jaw in all prosauropods and the majority of sauropods. The major exceptions were the horselike diplodocoid and titanosaur heads. In those, the teeth from the mid-jaw were lost, often to the degree that the teeth left over were limited to the front ends of the upper and lower jaws. The teeth were still numerous because they were slender, to the degree of being shaped like skinny pencils lined next to one another in densely packed in rows. Diplodocoid and titanosaur jaws tended to be more squared off than in other sauropods and, in a few cases, were very straight across. Set in sockets, the teeth were constantly replaced through life in the reptilian manner in order to prevent them from being worn down to nonfunctional, as can happen in plant-chewing mammals. Growing teeth pushed the worn examples off. Rates of tooth turnover ranged from modest to very rapid, as was the case with pencil teeth.

Among amphibians, tuataras, lizards, and snakes, the teeth tend to be set close to one another along fairly sharp-rimmed jaws, with the upper teeth always outside those on the lower jaw. The closed mouth is sealed, and the teeth are covered by lips that at least in most cases apparently cannot be pulled back to bare the teeth when the mouth opens. So even when the mouth is open, the teeth often remain entirely covered by lips, as well as gums; this is true of herbivorous iguanas as well as the flesh-eating monitor lizards. There seem, however, to be exceptions, such as geckos, hump-nosed lizards, and some snakes. These appear able to lift the lips to expose the teeth. If so, how they do so is not known. In any case, the saurischians very probably had

lips that sealed the mouth when closed—a sauropod skull appears to have a patch of lip tissue preserved astride some teeth. Thus, the artistic tendency to show teeth entirely exposed in saurischians looks like an error as big as it is common. Being herbivores, the possibility that prosauropods and sauropods would snarl is, however, low. Some contend that the tooth covering of sauropods was so stiff that it constituted a beak, and that is probably not strictly correct.

Some theropods and the ornithischians evolved beaks, and it has been suggested that prosauropods had incipient beaks. In ornithischians and therizinosaurs, the beak was limited to the front of the mouth, but in some theropods and many birds, the beak displaces all the teeth. Beaked birds lack lips, and most do not have cheeks, either. Condors, however, have short mouths because the sides of their jaws are covered by elastic cheek tissues, which differ from the muscular cheeks that cover the side teeth in many mammals. The side teeth of herbivorous prosauropods, the first sauropods, and ornithischians tend to be inset from the side of the mouth; the surrounding spaces are smooth-surfaced, and the foramina that feed the soft tissue in the area of the mouth are reduced in number and enlarged in size, indicating that well-developed, probably elastic cheeks covered some or all of the side teeth. Most sauropods did not have cheeks covering the lateral teeth. However, the abbreviated jaws of diplodocoids and titanosaurs suggest that the mouth was short in side view, with most of the jaws covered by cheeks.

Not being lizards or snakes, sauropodomorphs lacked flickering tongues. They did have well-developed hyoids, suggesting that the tongues the throat bones supported were similarly developed, being strong and supple enough to readily manipulate plant materials to assist rapid swallowing. The front end of some sauropods' lower jaw was downturned; this probably anchored the beginning of a subtle throat pouch that merged with the underside of the neck.

With their eyes on the sides of their heads, sauropodomorphs had a wide spread of vision, approaching in some cases 360 degrees. Any overlapping of forward vision would have been limited. Bony, sclerotic eye rings that helped

Buriolestes

Plateosaurus

unnamed genus *youngi*

Yizhoousaurus

Diplodocus

Nigersaurus

Camarasaurus

Giraffatitan

Nemegtosaurus

cover. The situation may, however, be more complicated, reducing the reliability of this method. In living animals, the relationship between the orientation of the semicircular canals and the normal carriage of the head is not all that uniform even between individuals of a species. That animals position their heads in different ways depending on what they are doing does not help. Giraffes feed with the head pointing straight down when browsing on low shrubs, or horizontally, or straight up when reaching as high as possible, so the orientation of the semicircular canals is not necessarily informative. The semicircular canals of at least some prosauropods seem to show that they typically held the nose tilted somewhat upward, an odd pose not normal in large herbivores. It seems that the posture of the semicircular canals is determined at least in part by the orientation of the braincase with the rest of the skull and does not reflect the orientation of the head as well as has been thought.

Sauropodomorph necks were rather short in early prosauropods, substantially long in the rest of the group and some sauropods, and up to ultra elongated in sauropods, with the longest-known neck in the area of 16 m (50 ft), eight times longer than those of giraffes. In tall sauropods, their towering height was mostly in the neck, with the forelimbs constituting only a quarter of the total in some mamenchisaurs. In the atypical atlasaurs, the arms were the primary height component. While giraffes have just the seven neck vertebrae present in nearly all mammals, prosauropods and sauropods sported 10–19—in the dinosaurs, the first cervical, the atlas, is very small. There has been a tendency to make sauropodomorph necks too short by placing the shoulder girdle too far forward; this is especially a problem of older skeletal mounts. Prosauropod neck vertebrae were fairly standardized, being moderately long, shallow, and narrow. Those of sauropods were highly divergent in all dimensions—some being very elongated—and in the side-view shapes of the neural spines. Longer necks with more vertebrae tend to be more flexible. Sauropods further enhanced both neck strength and mobility with ball-and-socket joints between the centra, a feature paralleled in giraffes. Birds have further special adaptations that make their necks exceptionally mobile, including in some cases extra large numbers of cervicals—up to 26—plus non-overlapping neck ribs. Most sauropodomorphs had the overlapping neck ribs more typical of archosaurs, a feature retained in crocodilians. In the slender-necked prosauropods and sauropods that retained rib overlap, the posterior rib was a very long and slender tapering rod that articulated with the next rib via a sliding articulation—after death, these often sprung apart. At a given location on the series, up to four rods might lie atop one another. Presumably, the overlap strengthened the neck without preventing flexion, the slim ribs being bendable. The rods were operated by

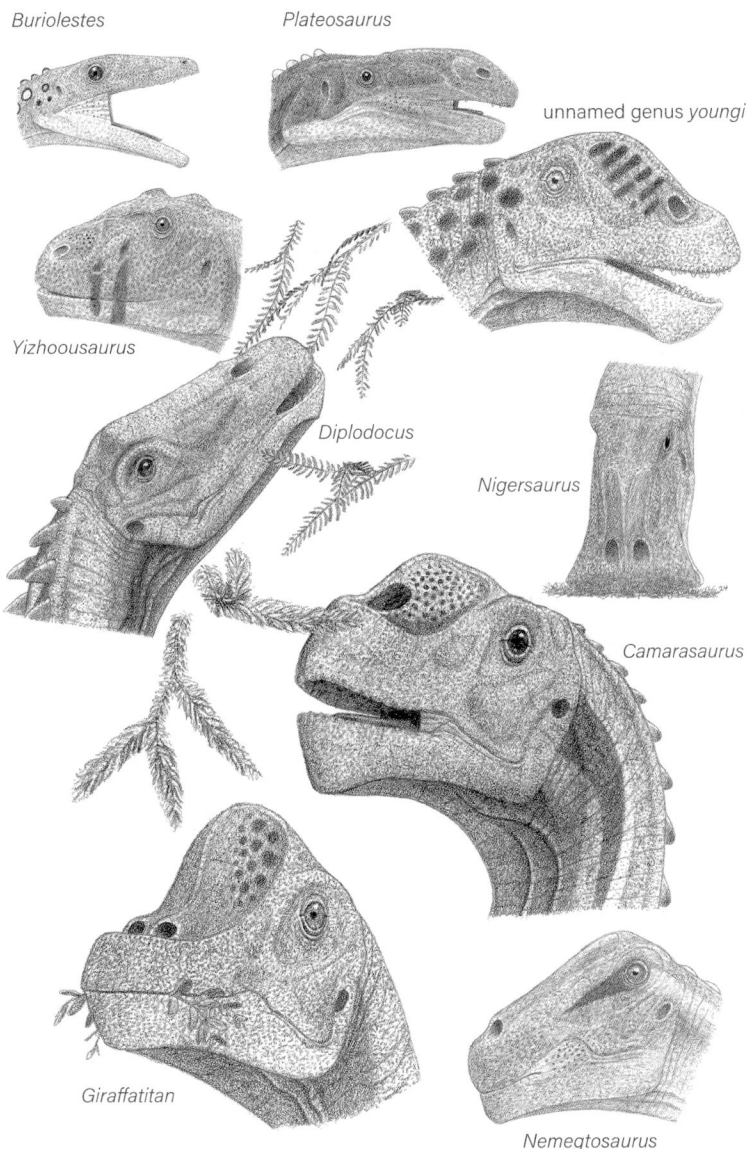

support the large eyes often show the actual size of the eye, both in total and indirectly, in that the diameter of the inner ring tends to closely match the area of the visible eye when the eyelids are open. Sauropodomorphs had large eyes, although relative eye size decreased somewhat as the creatures got bigger both during growth and between species. Prosauropod and sauropod pupils were very probably circular at all dilations; slits are limited to predators that need to cope with major variations in light levels. The eyes of birds and reptiles are protected by both lids and a nictitating membrane, so the same was presumably true in sauropodomorphs.

The outer ear is a deep, rather small depression between the quadrate and jaw-closing muscles at the back of the head. The eardrum was set in the depression and was connected to the inner ear by a simple stapes rod. The orientation of the semicircular canals of the inner ear is being used to determine the posture of dinosaur heads. For example, short-necked diplodocoid heads pointed straight down, according to this method, implying that they grazed ground muscles farther aft on the neck and front of the trunk. The overlap was lost within the diplodocoids, further increasing their neck flexibility in all directions. In most sauropods, the lower portions of the neck ribs are set well below the centra bottoms, a feature specific to group. The purposes of this unusual placement are not well understood, there being no modern analog. Just how much the neck of given sauropodomorphs could be flexed is difficult to determine because the cartilage intrajoint and joint capsules are long gone. In any case, the "neutral" posture that articulating the bones seems to produce sometimes differs depending on what portions of the joints between the vertebrae are used as the determining factor. Nor do the neck vertebrae of different giraffe individuals articulate consistently—they can range from arcing strongly downward in juveniles to being strongly erect in adults. This reflects the differing thickness of the cartilage pads between the vertebrae and demonstrates that cartilage as well as bones must be present to articulate necks properly. That is a major problem because cartilage rarely fossilizes. In any case, living animals are not particularly prone to holding their neck in the osteological neutral posture all that much, if at all. Ratites and giraffes hold their neck at different angles depending on what they are doing, including when they are not doing anything in particular. If anything, animals tend to hold their neck more erect than the neutral articulations suggest, if for no other reason than doing so helps improve the view of the world around them. In some cases, living animals can flex their neck in ways that are not in accord with their dry neck bones, such as camels touching their back with the top of their head, and ungulates nibbling their flanks.

The necks of many dinosaurs tended to articulate in a birdlike S curve ranging from very pronounced to gentle. The latter appears to have been true of prosauropods. The posture and function of the long necks of sauropods became highly contentious. Some researchers have proposed a simplistic model in which the necks of all sauropods were held nearly straight and horizontally and, in a number of cases, could not be raised much above shoulder level. To this add that elevating the brain much more than giraffes do requires even more extraordinary blood pressures that seem beyond plausibility. And the taller an animal is, the greater the pressure in the feet, to an extreme level in animals that were standing many stories high. If the 3 m (10 ft) limitation on the brain-heart height differential is correct, then the neck vertebrae of all large sauropods should have articulated with one another in a manner that precluded them from elevating much above horizontal. And there should have been no other means of raising the neck to a high angle.

An inability to raise the head much above shoulder level when standing on all fours appears true of one group, the short-necked diplodocoids. Otherwise, all large sauropods could loft the brain many meters above the heart. Many sauropod necks that have been restored in a straight line show obvious misarticulations or are based on vertebrae that are too distorted and/or incomplete to be reliably articulated. And there is the general absence of intervertebral cartilage. The well-preserved neck and trunk vertebrae of some prosauropods, and those of the tail base of others, are so strongly wedge-shaped that they describe impossibly strong downward arcs; in life, cartilage straightened that out. In many fossil skeletons, the vertebrae are found jammed tightly together, probably because the intervening cartilage disks dried out after death and pulled the bones together. In some articulated sauropod skeletons, the vertebrae are still separated by the substantial gap that was filled by cartilage. The only example in which the cartilage between the vertebrae was preserved in a sauropod neck is two neck-base vertebrae of a large, old camarasaur that had fused before death. Contrary to the prediction based on consistently horizontal-necked sauropods, the vertebrae are flexed upward as though the neck base was normally flexed upward and the head was held well above shoulder level. Because sauropod necks had so many vertebrae, just 10 degrees of upward flexion between each pair allowed most sauropods to raise most of the neck nearly vertically, with the head far above shoulder level. Some of the longer-necked sauropods could arc their necks up and back until the head was directly above the trunk. Diplodocoids could not do that, but the diplodocids with long necks could still elevate their heads by putting the entire neck into a strong U upcurve. There is no reason to assume that living sauropods had specific neutral neck postures or could not bend their neck more and normally hold it higher than the bone articulations may seem to indicate, as is the common case in living animals. Most researchers favor the probability that many sauropods regularly held their head high even when quadrupedal. That high vertical reach was almost

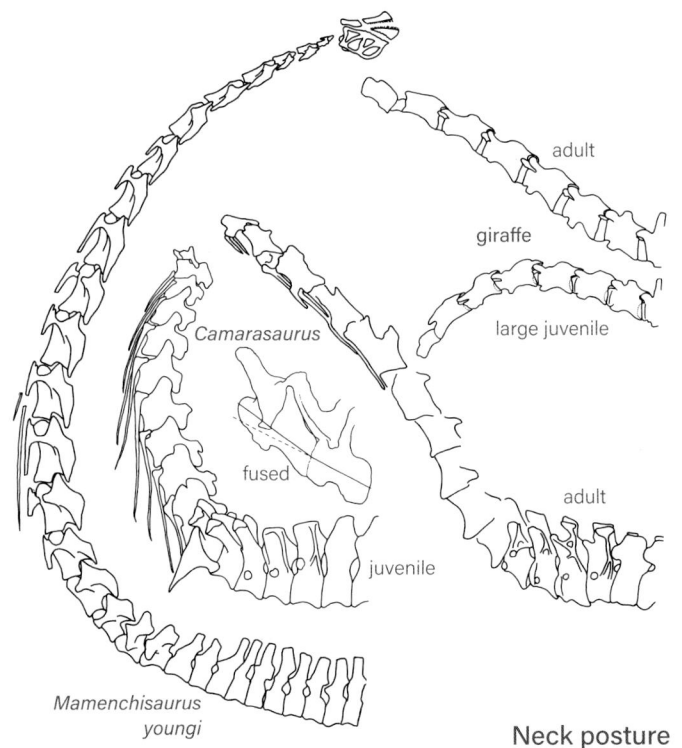

Neck posture

always achieved in sauropod clades largely by elongation of the neck over the forelimbs—the latter minimizing pressure issues at the heart, but not at the feet—the early atlasaurs being the one known exception, indicates that the pressure problem was not as much of an issue as it may appear to have been.

Pretensioned dorsal ligaments and tendons hold up necks without muscular effort, so pulling necks down requires effort. Prosauropods and short-armed sauropods with horizontal neck bases would have been able to reach the ground to feed and drink with little effort. Tall-armed sauropods with more erect bases may have had to pull the lower neck muscles to get their heads down enough to drink and spread their forelimbs to do it, like giraffes. Whether the arms were spread fore and aft like most drinking ungulates, or out to the side as do giraffes and okapis, is not known—note that drinking giraffes have to splay their arms not because of their long necks, but because their necks are not long enough to compensate for their tall forelimbs. Long-necked birds—ratities, swans, flamingos, and so forth—are not particularly good models for the breadth and depth of sauropod neck muscles relative to the vertebrae, the living dinosaurs' cervical series being very flexible, slender, and comparatively small. The big necks of giraffes are more informative. And they are not massively muscled, especially on their sides and bottoms, despite having solid vertebrae that have to support a heavy head. Holding up relatively smaller heads and being pneumatic, sauropodomorph necks would not have been massively muscled, either. In the slender-necked dinosaurs, the bulge of each neck vertebra should have been visible on the sides of the neck, as it is in giraffe necks when viewed from the front—the speckled pattern of their fur hides that otherwise-obvious feature in side view. The upper muscles that hold the head and neck up are markedly stronger than those that lower them, and the former formed a subtle bulge contour over the latter that ran along the sides of the upper end of sauropodomorph necks. The topography of these beasts was complex, not smooth and simple as sometimes illustrated. The necks of giraffes are quite narrow side to side. Those of sauropodomorphs are at least somewhat broader, particularly in sauropods, some of which are very wide. With researchers so focused on vertical action, what the width of sauropod necks means relating to their lateral function has not been explored in detail. It may reflect a greater ability of the dinosaurs to swing their necks to the side when being held erect, for feeding and/or intraspecific combat. The prominence of the cervical ribs may play a role in that. The presence of

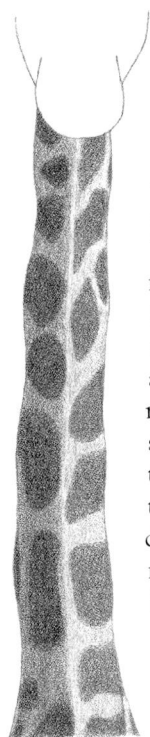

Sauropod neck cross-section

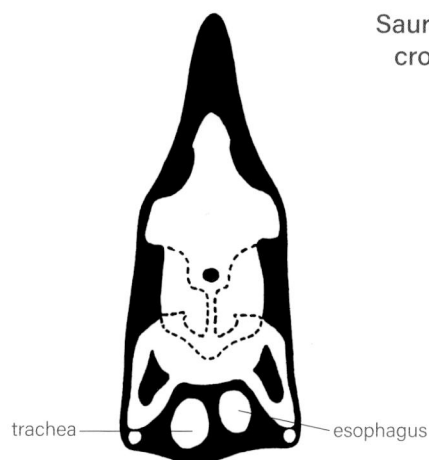

trachea — — esophagus

Giraffe neck

extensive lateral air sacs may also play a role in sauropodomorph neck breadth.

In some sauropods, tall shoulder spines indicate that a fairly deep set of nuchal ligaments and tendons—the latter operated by muscles anchored on the spines—helped support the neck, similar to some ungulate mammals, giraffes among them. In a number of other sauropods, the neck and shoulder neural spines were doubled in order to improve neck support and mobility. How the upper neck muscles, tendons, and ligaments were placed and operated in these dinosaurs with their unusual twin spines is difficult at best to determine. Because the trachea, esophagus, and cervical blood vessels of sauropods should have been tucked up between their downward-projecting cervical ribs in most cases, the bottom of their necks should have been fairly flat, unlike giraffe necks, which lack well-developed neck ribs. Being between the neck ribs would have kept the trachea and esophagus from flopping around the cervical series, which seems to happen in long-necked birds. At the juncture of the neck and shoulders, the absence of large underneck muscles and the prominent projection of the lower shoulder and sternal elements indicate there would have been a strong incurve at the neck-trunk transition, as there is under the base of giraffe necks.

Combining the dorsals and sacrals, prosauropods and sauropods had 15–17 trunk vertebrae. Sauropods had the fewer, articulated either in a nearly straight line or a subtle, dorsally convex arch. Prosauropods had three hip vertebrae, sauropods four or usually five, sometimes six or seven, to support their much larger pelvic complexes. The prosauropods' elongated dorsal vertebrae and shallow trunks indicate they could flex a fair degree. The trunk muscles should have been fairly well developed to achieve this flexion. The far taller trunk vertebrae of sauropods, the nature of the vertebral articulations, the in many cases partly ossified interspinal tendons, and deeper rib cages indicate that sauropods had stiffer backs, although their trunk vertebrae were not normally fused the way they often are in birds. Because sauropod trunk vertebrae and ribs formed a short, fairly rigid body, with the shoulder and hip girdles close together, the trunk musculature would have been rather light, like that of birds. The ribs are not visible in healthy elephants, but they usually are in rhino species, so this is optional in sauropodomorph illustrations. Also visible would have been the contours of the edge of the ilia and the tips of pubes and ischia. As in most vertebrates, including lizards, crocodilians, and birds, the chest ribs are swept back in articulated sauropodomorph fossil skeletons; they were not vertical as in many mammals. The chest is fairly narrow to allow the shoulder blades to work alongside them. The belly ribs tend to be more vertical in side view, but this condition is somewhat variable.

Unlike carnivores and more so than omnivores, herbivores need capacious abdominal tracts to break down plant materials. Belly capacity was substantial but modest in the shallow-bodied, narrow-hipped prosauropods. Larger, deeper pelves and posterior rib cages show that sauropods had larger, more complex plant-processing arrangements. Those of diplodocids were fairly narrow, albeit deep. Camarasaurs, brachiosaurs, and most of all titanosaurs evolved yet more commodious bellies, increasingly flaring out the posterior ribs and the forward ilial plates far out to the sides—so much so that the forward-most section of the ilium was primarily for abdominal rather than upper thigh muscle support, and the latter would have been hollowed out a modest amount. Only broad-beamed derived therizinosaurs matched them, and ankylosaurs exceeded them in this respect.

Prosauropods retained gastralia, a series of flexible bony rods in the skin of the belly. Each segment of the gastralia was made of multiple pieces. This may have been necessary in prosauropods because they flexed their trunks while running on all fours. These structures were scarce at best in sauropods. Why they were lost is not clear. Had they been retained, the cuirass could have helped support their big bellies while providing some protection against abdominal attack.

Consisting of 50–80 or so vertebrae that were not braced by stiff, ossified ligaments or tendons, sauropodomorph tails were flexible. Medium long and deep, prosauropod tails were not especially bendable. With more caudals and generally shallower at mid-length, sauropod tails are more elastic, especially toward their tips. Most flexible were the whip tails of diplodocoids, diplodocids especially, and titanosaurs. A ball-and-socket joint at the base of titanosaur tails allowed the majority of the structure to be arced directly over the back. The tails appear to have been gently arched up a little as they emerged from the hips in many cases. In a few examples, the tail base was strongly tilted dorsally. The base of the tail was well muscled, both with muscles that operated the tail and with the very large caudofemoralis, which inserted on the back of the femur. These muscles were very powerful leg retractor propulsive units, especially in those sauropodomorphs with deep tails behind the hips—the bottom of the caudofemoralis formed a prominent bulge as it went into the thigh. The

whips are so slender that they were probably not muscled or even operated by tendons, and probably hung semilimp until swung into action by the rest of the tail. Among a few sauropods, the last tail elements were fused into clubs or rods. How often this was true is not entirely certain because tail tips are absent for many species. The tail was held off the ground by pretensioned tendons and ligaments in the same manner as the neck, and pulling it down to the ground likewise required work by the ventral muscles.

Turning to limb form and function, all sauropods were long-legged in the elephant manner; none had the short limbs of low-feeding rhinos or hippos. Among larger sauropods, the combination of their long limbs and sheer size would have allowed a person to walk beneath the belly without hitting their head. With the pelvis directly attached to the sacral vertebrae, the position of the hindlimbs is obvious. Because the shoulder girdle is not directly attached to the vertebral column, where and how the forelimb attaches to the trunk is not so tightly fixed, requiring reconstruction of the position of the arm. In most tetrapods, the shoulder blade, which usually consists of the scapula-coracoid—mammals are unusual in having severely reduced or lost the coracoid—is subvertical; such was true of prosauropods and sauropods. Among archosaurs, only pterosaurs and birds have horizontal scapula blades as part of their specialized flying apparatus, and restorations showing the terrestrial sauropodomorphs with such are inaccurate. Among the two-legged theropods, the cofused furcula locks the shoulder girdle into a rigid structure for purposes of predation and flight. Many tetrapods have mobile shoulder girdles that, among quadrupeds, increase the stride of the forelimbs. Lacking a rigid furcula, the four-legged sauropodomorphs should have had mobile scapula-coracoids. There were clavicles in prosauropods and some sauropods, but unfused clavicles are mobile and do not prevent considerable shoulder-girdle motion in lizards and many mammals, humans included. This is all the likelier in that the shoulder blades of four-legged dinosaurs resemble those of chameleons, which have highly mobile shoulders. The coracoids articulated with fore- and aft-sliding grooves along the sides of a narrow midline cartilage (rarely partly ossified in dinosaurs), which in turn attached to paired, kidney-bean-shaped sternal plates that were often very large in sauropods, especially the broad-bodied examples. These in turn were connected to the rib cage by sternal ribs. Those were typically cartilaginous, which is another mystery in that their being bony would seem to have improved the strength of the forelimb-to-trunk connection, and in a few sauropods, they appear to have been ossified. While the shoulder girdle should not be placed too far forward, it should not be put too far aft, either—the shoulder joint, with its broad-beamed humerus head, was always forward of the first chest ribs, not astride them, as in many four-legged mammals.

All dinosaurs except for flying birds have been hindlimb dominant in that, even in the quadrupeds, the legs always bear more of the animal's weight than the arms. We know

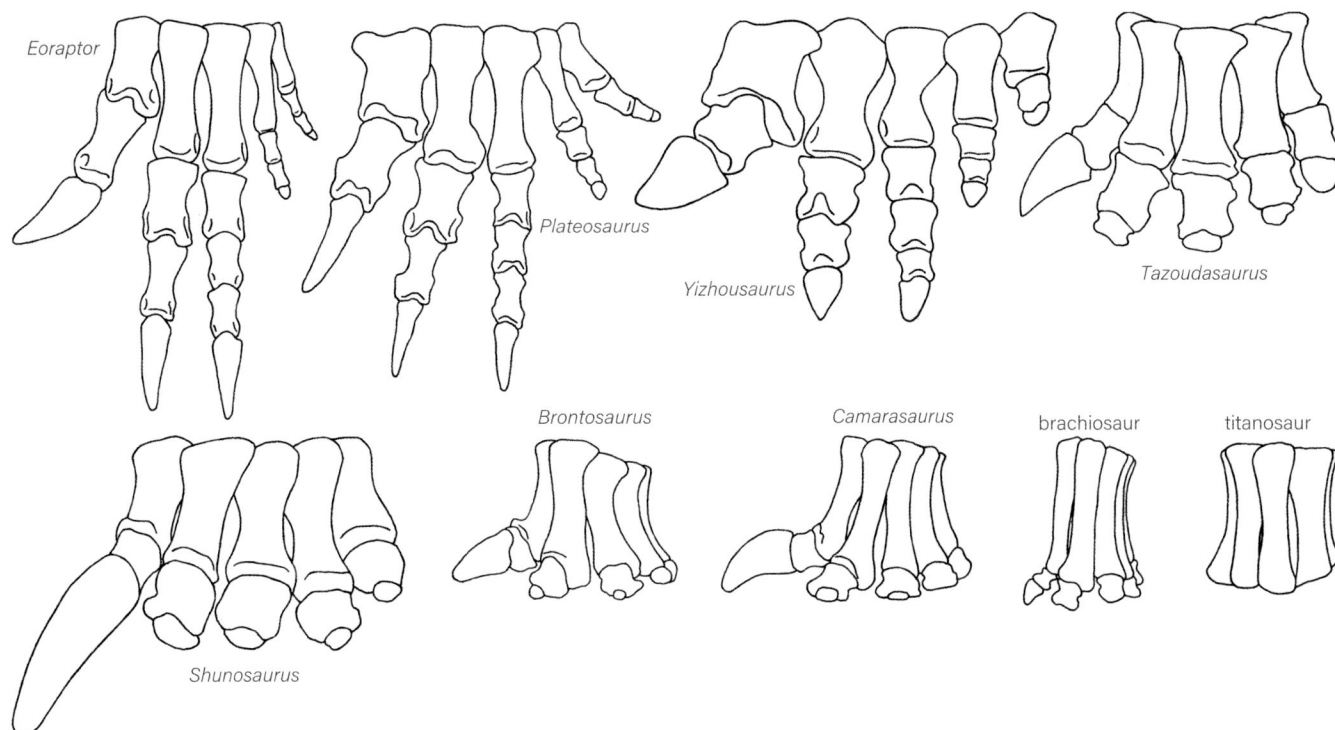

Eoraptor   *Plateosaurus*   *Yizhousaurus*   *Tazoudasaurus*

Brontosaurus   *Camarasaurus*   brachiosaur   titanosaur

Shunosaurus

this because the femur is always thicker-shafted than the humerus. And the hindfeet are always larger than the hands. This is true even of the longest-armed brachiosaur sauropods. The reason for this is that the heavy tail always added mass load to the hindlegs. Not having much in the way of tails while having big heads leaves many mammals forelimb dominant, like ungulates and elephants, whose humerus is thicker than the femur and whose forefeet are larger than the hindfeet—flying birds are forelimb dominant because their wings are their main propulsive organs. In tune with this scheme, dinosaur hands are always very different from the feet. In many quadrupedal mammals, the hands and feet are nearly identical, as per ungulates, elephants, carnivores, most primates, etc. The five-fingered prosauropod hand was quite short and broad, with supple fingers that became smaller progressing laterally. The thumb bore an extra large, strong hook claw; the outer two fingers were clawless. The foot was much longer, with four functioning toes, the inner shortest and with the biggest claw, the third toe the longest. Trackways show that the digits bore

light padding similar to big birds. Both appendages were digitigrade, with the wrists and ankles clear of the ground, as per birds and cats and dogs, in contrast to plantigrades, such as humans and bears. There is little variability in prosauropod hands. Sauropod extremities were very different from both prosauropods and each other. The hindfeet were similar to those of elephants. Very short, broad, and based on a big, round, elastic pad that flattens out some when pressed onto the ground, the foot was fixed close to inflexible on the shank. The biggest difference with elephants was that sauropod feet sported three and sometimes four round-tipped, banana-shaped claws, directed somewhat to the sides, decreasing in size laterally like the toes that they tipped. Once it evolved, the sauropod foot did not change all that much over the span of the group. The sauropod hand was radically different. It was made up of five upper hand elements, forming a vertical, columnar, tightly bound, lunate arcade. Fingers were very abbreviated and became increasingly reduced until they were entirely missing in titanosaurs. When present, the short thumb carried a blunt

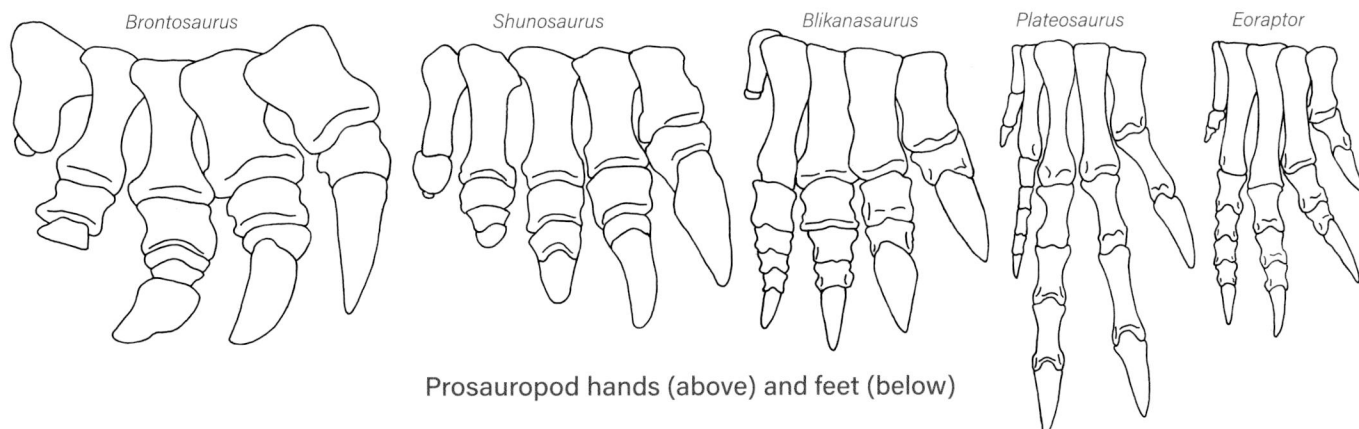

Brontosaurus   *Shunosaurus*   *Blikanasaurus*   *Plateosaurus*   *Eoraptor*

Prosauropod hands (above) and feet (below)

claw, sometimes large, sometimes not, and missing in most titanosaurs. This rather ungulate-like scheme lacked major padding. The trackways record a hollow lunate shape with the hollow directed aft and usually somewhat inward. Sauropod hands and feet were almost unguligrade, in that they were perched more toward the tips of the toes than usual in flatter-toed dinosaurs—the fingerless titanosaurs were in a sense beyond unguligrade, being uniquely metacarpus-grade among vertebrates. Presumably, this more vertical orientation improved weight bearing in these behemoths.

Many mammals can readily rotate their lower arm along its long axis to swivel the orientation of the hand, but others cannot do so as easily, if at all. Dinosaurs were more like the latter, so their palms usually faced partly or strongly inward. This is recorded in most prosauropod and sauropod trackways, but in some cases, the medial orientation is not strong. In order to allow the sauropod hand to be arranged in as much a fore-and-aft manner as possible, the elbow articulation was modified to place the head of the radius more internally than usual. The prosauropod wrist was very flexible, consisting of numerous carpals that could readily move relative to one another. Sauropod carpals seem much simpler, which can be taken to indicate that the wrist was not very mobile. But wrist flexion is important in striding animals to help clear the forefoot from the ground during the recovery stroke when on the move—the very slow tortoises have fixed wrists. Ungulates have fairly simple carpal blocks that partly disarticulate when the wrist is flexed. And some better-ossified sauropod carpals hint that their wrists were more complex than realized.

The vast majority of fossil trackways were laid down on wet, soft muds and sands along watercourses, much fewer on dry sands, such as dune faces. From personal experience, doing either requires considerably more work than progressing on firm substrates to the point of being exhausting. According to their trackways, prosauropods were usually bipedal—one has to be careful about this, since it is possible for hindprints to cover over foreprints, as is normal with elephants—but the dinosaurs also progressed on all fours. Some prosauropods had arms too short to walk at a fast quadrupedal clip; others had arms too long not to, while others would have been adept at both gaits. All sauropod forelimbs were good for progressing on all fours, and the animals normally did so. In both prosauropod and sauropod trackways, the hindfeet were typically placed behind the handprints as their print makers strode across the long-ago Mesozoic terrain.

Trackways affirm that hands and feet fell along a generally narrow gauge, as expected with limbs operating in a near-vertical fore-and-aft plane. It is difficult to restore the precise posture of dinosaur limbs because in life the joints were formed by thick cartilage pads similar to those found on store-bought chickens, which are immature. That dinosaurs normally retained extensive cartilage joints throughout their entire lives, no matter how fast or big they became, is a poorly understood difference between

them and adult birds and mammals, which have well-ossified limb joints. The manner in which dinosaurs grew up and matured may explain the divergence. In terms of locomotory performance, it does not seem to have done dinosaurs any more harm than it does big running birds that still have cartilage joints when fully grown and running fast but not yet mature. Instead, it may have had advantages in distributing weight and stress loads and may be a reason sauropods were able to get so large. Despite the loss of detailed morphology of the cartilage joints, some basics about their posture and function can be determined. The shoulder joints faced downward and backward so that the arm could swing directly below the shoulder joints during locomotion, and the cylindrical hip joints forced the legs to work below the hips. But this does not mean that erect limbs worked in simple, entirely vertical fore-and-aft planes. Sauropodomorph elbows and knees, for instance, were bowed somewhat outward to clear the body, a feature common to other dinosaurs and mammals as well. Trackways show that unlike the forefeet of mammals, which are often near or under the body midline when walking, prosauropod and sauropod hands were usually separated by at least two hand widths; the hands were rarely placed closer to the midline than the feet and were often farther from the midline than the latter. This was because the arms were oriented so that the hands were either directly beneath the shoulder joints or a little farther apart, rather than sloping down and inward as typical of mammals. The hindfeet of sauropodomorphs often did fall on the midline. In sauropods, however, the gauge of the trackways differs. In most, it was on the narrow side. But in brachiosaurs, it was prone to be broader, and in titanosaurs even more so. This "wide gauge" posture was related to the extra wide bellies and hips of these broad-beamed dinosaurs. But even titanosaur feet were never separated by much more than the width of a single hindprint—some rearview postural illustrations of titanosaurs show the hindlegs spread too far apart.

Prosauropods had, like most terrestrial tetrapods, strongly flexed shoulder, elbow, hip, knee, and ankle joints that provided the springlike limb action needed to achieve a full run in which all feet were off the ground at some point in each complete step cycle. In addition, the ankle remained highly flexible, allowing the long foot to push the dinosaur into a ballistic stride. It is possible that the long, flexible, trunked prosauropods with longer arms achieved a bounding gait like galloping crocodilians. Even so, like crocodilians, prosauropods were not gracilely proportioned, swift sprinters, and top speeds should have been inferior to those achieved by top human athletes, probably in the area of 30–40 km/h (20–25 mph).

Sauropods evolved elephantine, columnar, straighter-jointed limbs. The same also appeared in advanced stegosaurs. Their shoulder joints face more downward than in other four-legged dinosaurs, such as ceratopsids, and the elbow joint at the end of the humerus is the same. In sauropods and stegosaurs, the lower portion of the scapula

Prosauropod and sauropod trackways

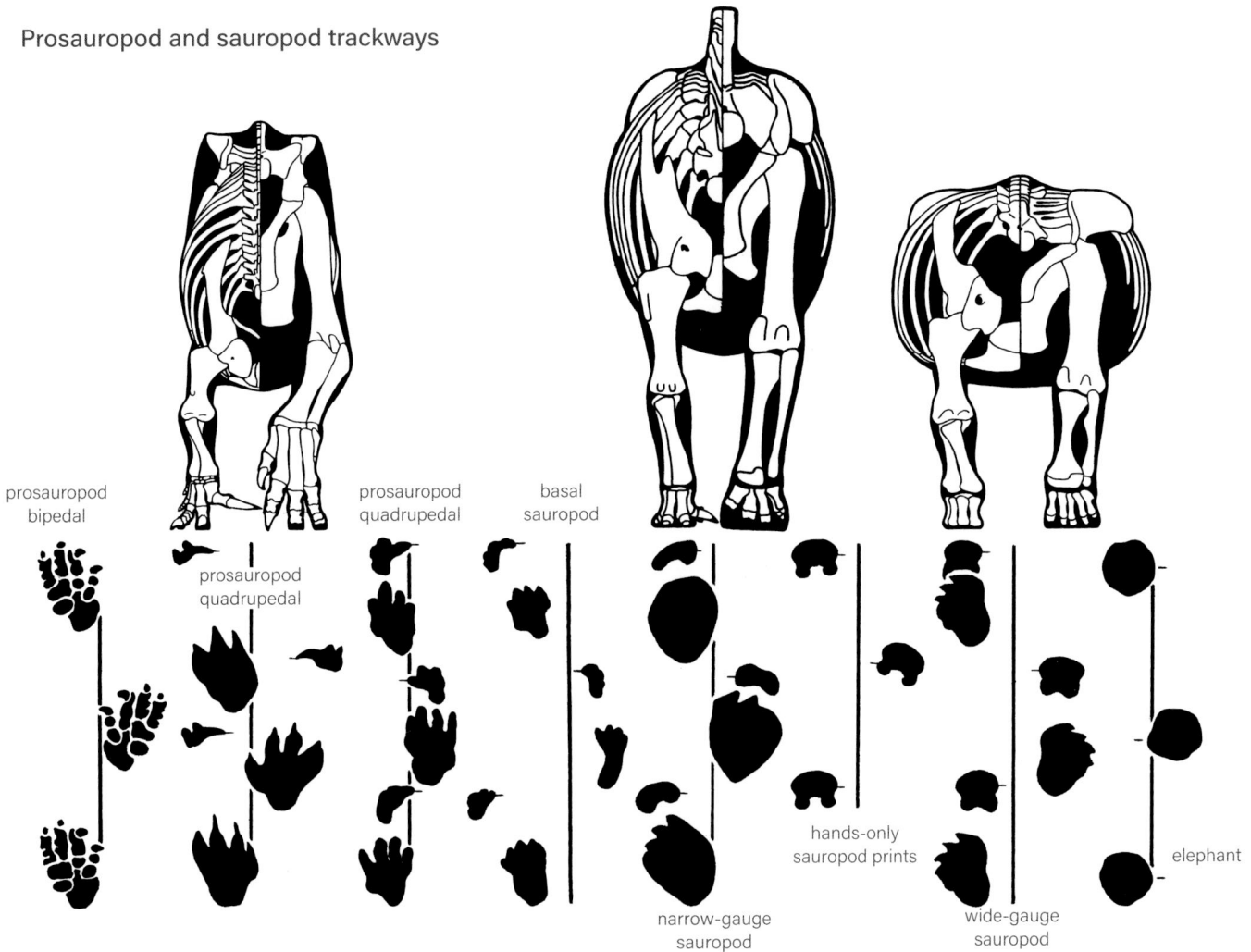

prosauropod
bipedal

prosauropod
quadrupedal

prosauropod
quadrupedal

basal
sauropod

hands-only
sauropod prints

elephant

narrow-gauge
sauropod

wide-gauge
sauropod

has an unusual forward expansion of the plate. This was a location for arm protractors to anchor upon, and from there they could swing the unusually vertically oriented upper arms well forward. Because the upper arm was vertical, those of sauropods could be a long segment of the forelimb, like elephants. Only animals with short upper arms can walk on seriously bent elbows, and restorations showing sauropods doing so are in error. Elbow and knee flexion varied during the propulsive stroke, being least when at the beginning and end and most during the mid-stroke, to keep the gait as smooth as possible. Straightening out the elbow and knee at the end of the power stroke helped propel the animal forward. When standing, elephants can actually lock the elbow for maximum static support by forward flexing it a little bit; there is every reason to presume sauropods did the same. In most dinosaurs,

Plateosaurus galloping

Shoulder joint flexion

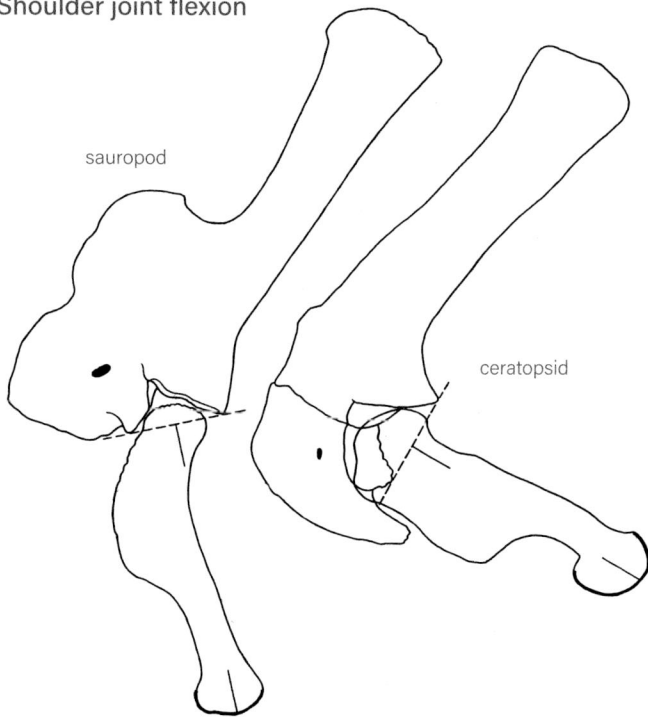

including many giants, the normally flexed knee could not be fully straightened because doing so disarticulated the outer femoral condyle from its contact with the groove formed by the proximal end of the fibula and tibia—this remains true in birds, as can be observed with the carcass of an uncooked chicken or turkey. In sauropods and stegosaurs, the upper end of the fibula was elongated a little, so that it continued to articulate with the condyle of the femur when the knee was entirely straight. The sauropod and stegosaur ankle was straight and had minimal mobility, like those of elephants and tortoises.

Because sauropods were so big-bodied, it might be tempting to imagine that their legs were laden with great, bulging body-builder muscles in order to bear their massive mass on land. But the similarly built elephants have modest limb muscles. The upper arm and the thigh are narrow. The shank muscles are especially unimpressive; hardly any fibers are on the front of the tibia, much as we can feel by hand on the front side of our own tibia. Elephant leg muscles are so modest because their vertically directed limb bones were posed to bear the weight load with little muscle work, and there was not a need to operate the nearly immobile foot. And elephants cannot run, so they do not need the power—conversely, their modest muscle are a reason they are slow. The same would have been true of sauropod limbs. The upper edge of the ilium was visible in these herbivores, in the same way that the pelvic bones of a cow can be seen under the skin. Because prosauropods were small-hipped, they did not have broad thigh muscles, either. At the knee, the tibia had a substantial forward-projecting cnemial crest, so there would have been a chicken-leg-like drumstick bundle of muscles on the upper shank that operated the feet via tendons. As usual with dinosaurs, the upper end of the sauropodomorph

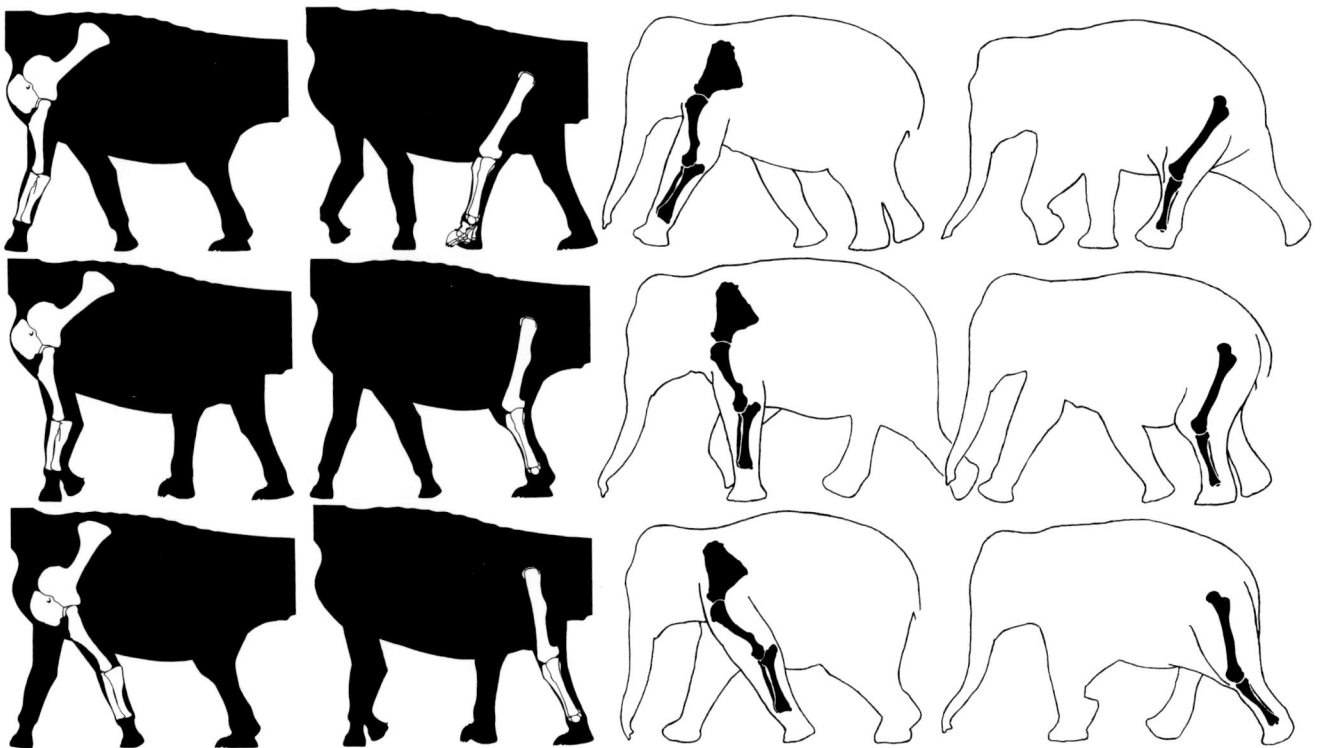

Fast ambling sauropod and elephant leg action

humerus bulged out a little, and the crest of the humerus formed a prominent contour along the upper front edge of the arm. The elbow joint and muscles around it formed a large bulge in front view; subtler bulges were at the wrists and ankles. The latissimus dorsi sheet on the flanks that helped pull back the arms likely formed a contour where it entered the back of the upper arm.

Being borne on elephantine-style arms and legs, sauropods would have had similar locomotary performance. Elephant races show they top out at an unimpressive 25 km/h (15 mph) or so; claims of faster speeds are spurious. Lacking the bouncy, flexed joints and long, mobile hindfeet, or the powerful limb muscles, needed to propel animals into a run with a suspended phase in which all feet are in air for a brief moment in the limb cycle, elephants are limited to an amble, a four-legged form of a walk-run in which at least one foot is always in contact with the ground. This is true regardless of size; juvenile elephants cannot move any faster than their parents. From when sauropods left their nests to when they died, they, too, were unable to achieve a true run with a suspended phase. Some calculations propose that gigantic sauropods were slower than elephants. But the longer stride length that comes from bigger dimensions should have allowed supersauropods to get to 25 km/h or so.

In principle, the top speed of extinct land creatures were recorded and can be determined by the measurements of the high-speed trackways they left behind. In practical terms, animals take the great majority of their steps while walking at a comfortable, sustainable walking pace. And the soft, mucky, slippery flats in which the trackways were usually preserved discouraged running. The speed at which a fossil trackway was laid down can be approximated—with emphasis on the approximated—by correlating the stride length of the trackway with the length of the articulated leg, from the hip joint to the base of the foot. The latter can be estimated from the length of the foot, which is four to one in a surprising array of animals, including prosauropods and sauropods, except that the ratio may have been five to one in some lithe diplodocines. Mammals and birds of all sizes tend to walk at speeds of about 3–7 km/h (average 3 mph). Note that squirrels will bound rapidly, halt for a moment, then bound some more, stop, and so on. Humans and their dogs typically move at a similar overall pace, as do elephants. This is because the cost of locomotion per given distance scales closely to available aerobic power as size increases, so being big does not provide a major advantage. Sauropod tracks show them walking 1–10 km/h (1–6 mph), broadly matching normal elephant paces. Fossil trackways recording speeds near maximum are rare because animals don't run very often, especially on wet, soft substrates, and none are known for sauropodomorphs.

Being bipedal to varying degrees, prosauropods could easily stand and slow walk with the body subvertical, using their heavy tails as props when not moving, to maximize their upward reach to crop desirable vegetation. Although they were big-armed, the hindlimb-dominant sauropods could all readily rear up, too—they had to do so for males to mount females. Notably, African savanna elephants rear up fairly often to feed on choice items, even though they are heavy-headed, forelimb dominants without substantial tails; for reasons not known, Asian elephants go bipedal in the wild much less often, if ever, despite their ability to do so in circus performances. The rearing ability of sauropods varied as well. Most appear not specialized to go bipedal to feed, although it may not have been a rare habit among the big-tailed dinosaurs. Diplodocoids were especially hindlimb dominant, being big-hipped, heavy-tailed, and short-bodied. To that add they had special sled-shaped chevrons beneath their tails like those of kangaroos, which use their tails as props. The group had the attributes that indicate they were specialized to rear up on legs and tail and stand tripodally. This was taken to an extreme in apatosaurs, with their especially massive pelves, tall hip vertebrae, and extra short trunks. With the pelvis titled up so strongly, the legs would have had trouble functioning properly, so the knees were probably somewhat flexed, as per rearing elephants. This was an immobile posture. A number of other sauropods also had the sled chevrons that indicate habitual tail-prop rearing. In some sauropods, the hips were retroverted and, along with the base of the tail, were flexed upward relative to the trunk vertebrae. This allowed the trunk to be held strongly pitched up while the hips and tail remained horizontal. This was best developed in camarasaurs, most of all *C. lewisi*. The therizinosaur theropods also had this peculiar arrangement. Their pelvic retroversion allowed the herbivorous bipeds to maximize vertical reach via an erect body while high browsing.

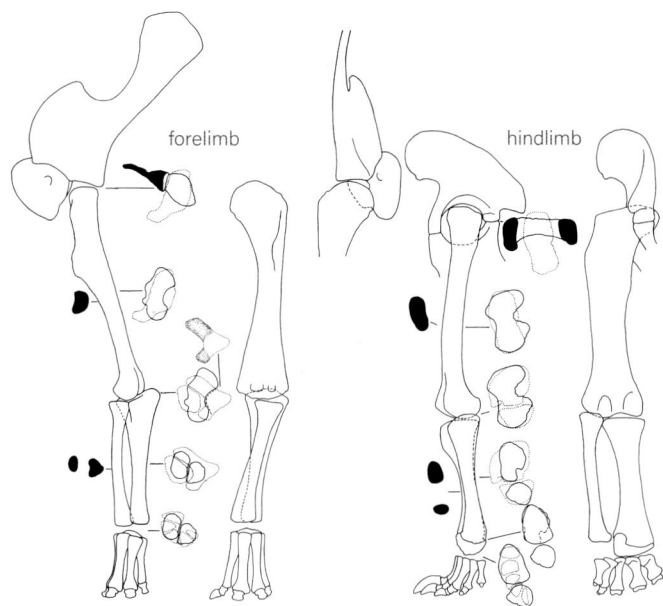

Sauropod limb articulation and posture

41

High-reach postures

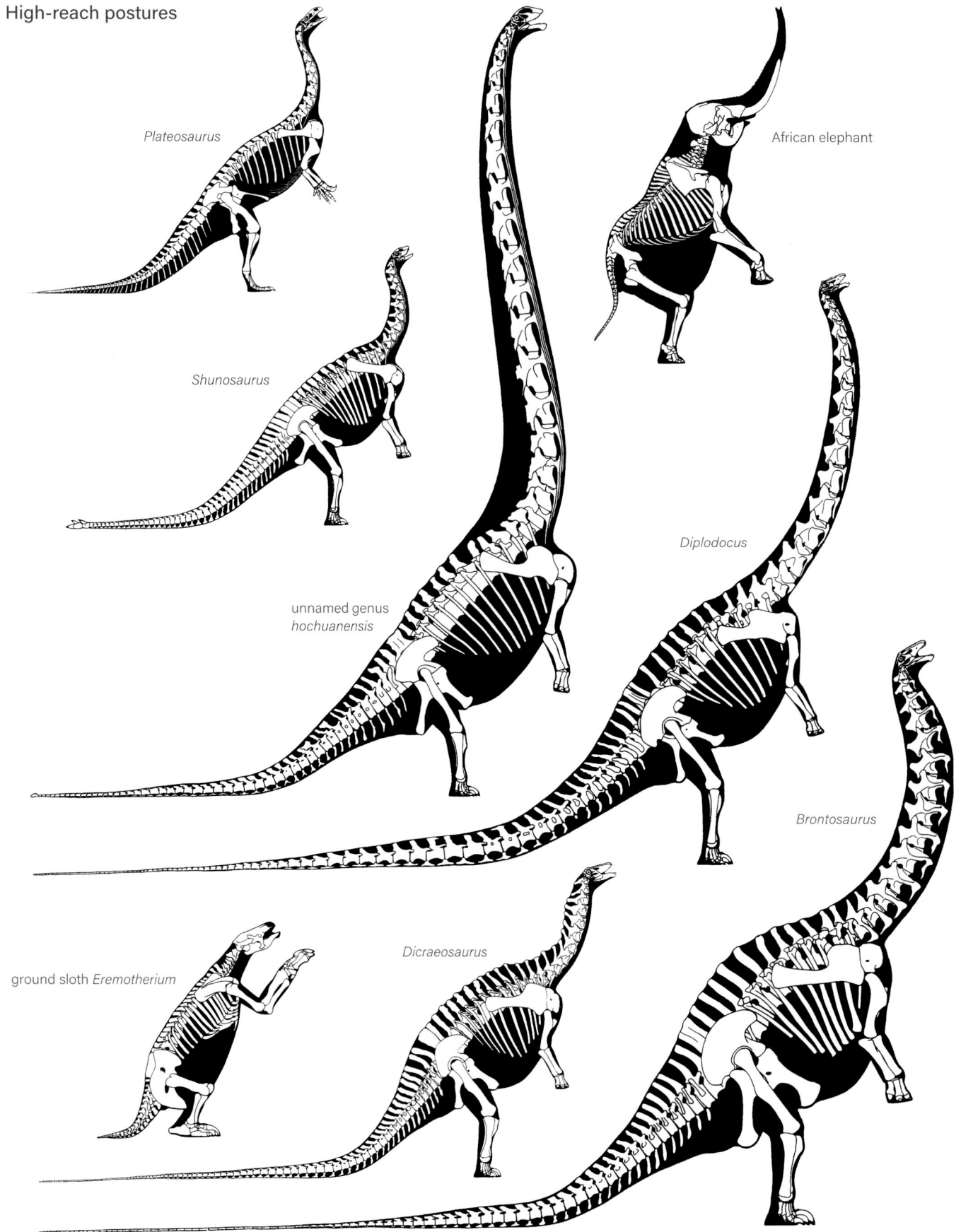

Plateosaurus

African elephant

Shunosaurus

unnamed genus
hochuanensis

Diplodocus

Brontosaurus

ground sloth *Eremotherium*

Dicraeosaurus

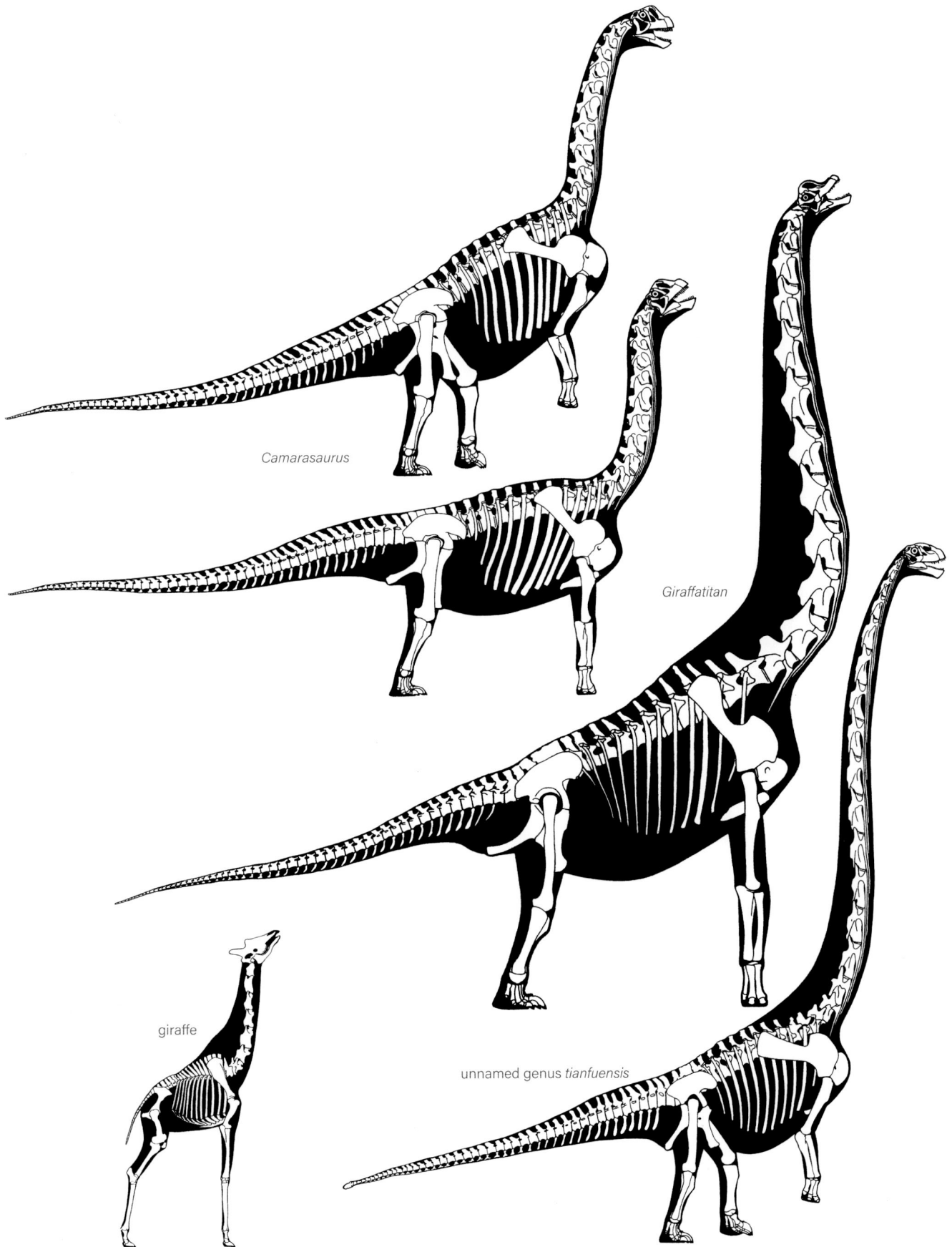

Camarasaurus

Giraffatitan

unnamed genus *tianfuensis*

giraffe

Camarasaurus, Barosaurus, and *Brontosaurus*

A popular conceit is that a super-sauropod shook the ground with each ponderous step even when slow walking. This is nonsense: elephants walk with essentially no sound on clear, solid soil. A herd of stampeding sauropods, on the other hand, would have been a thundering ground shaker.

When not on the move, how did the dinosaurs rest? This would have been easy for the modestly sized prosauropods. They could lie on their bellies with limbs tucked up beneath them, or on their sides. Matters were not so simple for ponderous sauropods. Elephants can lie on their sides, but the endless necks of many sauropods would seem to preclude them from doing so. When resting on their abdomens, the legs could not be tightly folded up, the same as with elephants.

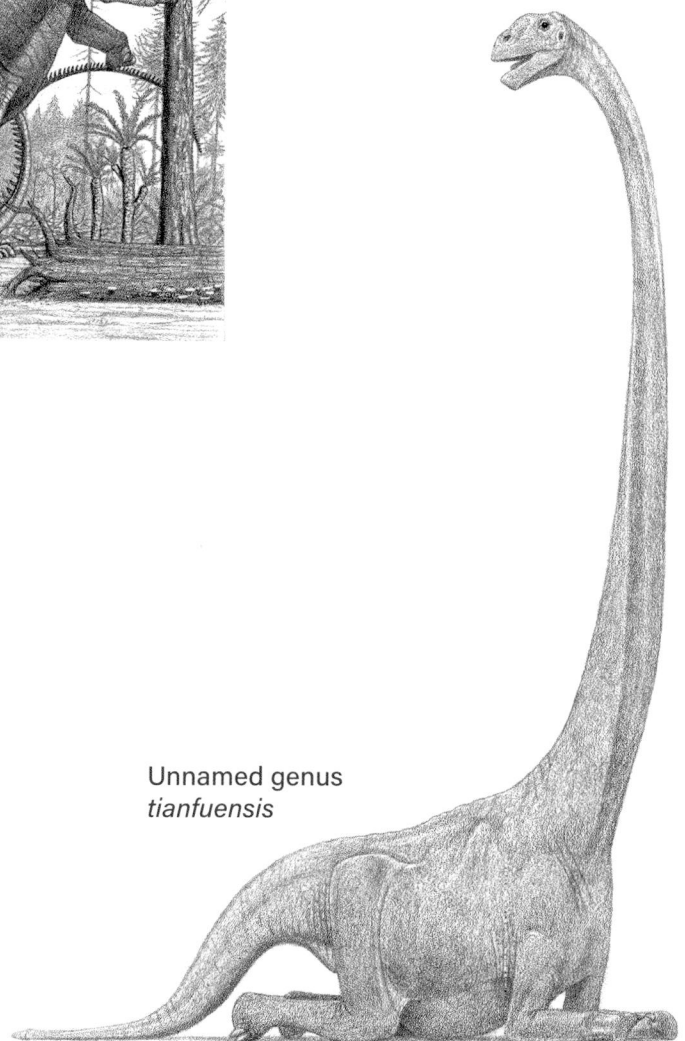

In camarasaurs, the slanted hips seem less than optimal because the backward pelvic tilt was sufficient to come close to hindering leg action when quadrupedal. The explanation is that, when rearing, the camarasaurs' hips were then horizontal to the ground, so the hindlegs could operate properly and the animals could slow walk on their two legs alone. Some other sauropods had somewhat retroverted hips that indicate they could also do a bipedal slow walk. Brachiosaurs, too, had the tilted-back pelvis; this was associated with their up-tilted trunk related to their very long arms. Brachiosaurs would seem least prone to rearing, but this does not mean they would not do it when it was the only means of reaching high-quality browse.

A few sauropod trackways consist of only hindprints. These could record occasional bipedalism. Or the big hindfeet were stepping onto the smaller foreprints—elephants normally overimprint this way. Another possibility is that hip-heavy sauropods were poling along the bottom with the hindlegs while swimming.

Unnamed genus
tianfuensis

SKIN AND COLOR

Most dinosaurs are known from their bones alone, but we know a substantial amount about dinosaur body coverings from a rapidly growing collection of fossils that record their integument. It has long been known that large, and some small, dinosaurs were covered with mosaic-patterned scales. These can be preserved as impressions in the sediments before the skin rotted away, but in numerous cases, traces of keratin are still preserved. Footprints sometimes preserved the shape of the foot pads. Lizard-like overlapping scales were not common among dinosaurs and have not been found among sauropodomorphs. No trace of a body covering is yet known for a prosauropod, except for some footprints that record the scales on the bottom of the feet. Among sauropods, there is a fair amount of material, much recently found and not yet detailed. None has been documented from the heads. Skin patterns are known for juveniles as well as adults.

As is common for dinosaur scales, those of mature sauropods are generally semihexagonal. Size is very variable, ranging from 5 to 30 mm (0.3–1.5 in). The dimensional difference can be seen in a small patch, although there is a possibility that scales higher up on the body tended to be larger than those on the underside. Sometimes a large scale is ringed by smaller ones, forming a rosette, another frequent dinosaur feature. Evidence of organization into larger patterns has not been seen in adults. The scales are fairly flat, but they are not smooth-surfaced. Unlike other dinosaurs, mature sauropod scales sport dense fields of small, beadlike, protruding papillae 1–3 mm in size (less than an eighth of an inch)—sauropod skin would have had a rather rough texture. Because sauropod scales were usually not large, they tend to disappear from visual resolution when viewed from a dozen feet or more.

Embryo sauropod scales are generally similar to those of their parents, albeit far smaller and sometimes more beadlike, and lack the papillae. There were rows of larger scales that may have run atop the backbone. Scale scraps from what may be a larger growing diplodocid include squarish and rectangular scales sometimes set in irregular rows, which may have followed the line of skin folds where limbs blended into the body.

For decades, sauropods were, ironically, restored like giant mammals in that their dorsal midlines were left unadorned—this even though midline frills were known in hadrosaurs, albeit not ceratopsids. The plain back art ceased when it became known that spiky frills like those of iguanas ran atop diplodocids, all the way to the tips of their whip tails. The midline scale rows of the embryos may have been the ontogenetic predecessor of the frills. It is generally presumed that frills were universal among sauropods and perhaps prosauropods, although this is not certain; some examples may have had partial frills.

Most sauropods were not armored. Titanosaurs bore substantial osteoderms up to 400 mm (1.3 ft), half that size or less being typical. Shapes were usually oval, also subcircular, teardrop, sometimes occasionally elongated. In profile, they were shallowly subtriangular. Sometimes there was a midline ridge that probably ran forward and aft. These have not been found in place, were not highly numerous on a given individual, and did not form a dense pavement, as they do on ankylosaurs. They may have been set in a couple of rows or more on the flanks, but that is speculative.

It cannot be overemphasized that illustrating sauropods with elephant-like wrinkled skin is entirely incorrect. Such a peculiar integument is limited to proboscideans—rhino skin, for instance, is not similarly wrinkled. Throat and neck wattles and similar soft-tissue display structures are plausible on prosauropods and sauropods. Some lizards have vertical skin folds around the shoulders, and such is possible in these dinosaurs. There is evidence that simple protofeathers evolved at or before the base of the dinosaurs. If so, they were lost at some point in sauropodomorphs, perhaps among the basal prosauropods. That issue awaits more fossil evidence.

Pigment capsules are allowing the colors of feathered dinosaurs to be determined to a fair extent, and such is now being done for some of the dinosaur skin fossils that incorporate original tissues. Those include a diplodocid that had a yellowish or ginger color. This indicates that sauropods were not a uniform gray in the manner of giant continental mammals, although it is quite possible that some of them were such. The hypothesis offered by some that the differing scale patterns on a particular dinosaur species correspond to differences in coloration is plausible, but some reptiles are uniformly colored regardless of variations in scales. Dinosaur scales were better suited than

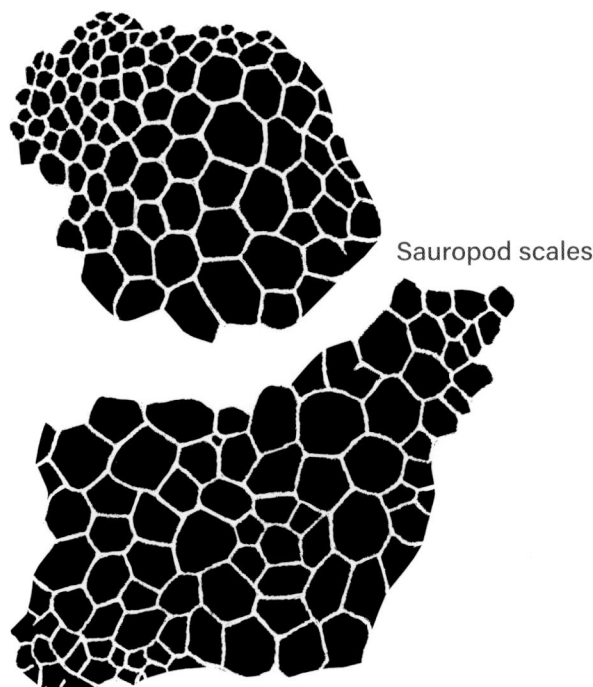

Sauropod scales

the dull gray, nonscaly skin of big mammals to carry bold and colorful patterns like those of reptiles, birds, tigers, and giraffes, and the color vision of dinosaurs may have encouraged the evolution of colors for display and camouflage. Sauropodomorphs adapted to living in forested areas may have been prone to using greens as stealth coloring; drier areas probably favored more browns and tans. Being dark-skinned could pose overheating problems for species living in hot climes; for those living at higher latitudes, dark may have been better. Small dinosaurs are the best candidates for bright and/or bold color patterns like those of many but not all small lizards and birds. Dinosaurs would be likelier to bear disruptive color patterns when very vulnerable juveniles than when adult. On yet another hand, because humans lack vision in the UV range, we miss seeing a lot of the coloration of many animals, so a number of reptiles and especially birds that look drab to us—including females that look bland and much the same—feature dramatic UV color patterns, often for sexual purposes. Archosaurs of all sizes may have used specific color displays for intraspecific communication or for startling predators. Crests, frills, and taller neural spines would be natural bases for vivid, even iridescent, display colors, especially in the breeding season. Because dinosaur eyes were bird- or reptilelike, not mammal-like, they lacked white surrounding the iris. Dinosaur eyes may have been solid black or brightly colored, like those of many reptiles and birds.

RESPIRATION AND CIRCULATION

The hearts of turtles, lizards, and snakes are three-chambered organs incapable of generating high blood pressures. The lungs, although large, are internally simple structures with limited ability to absorb oxygen and exhaust carbon dioxide and are operated by rib action. Even so, at least some lizards apparently have unidirectional airflow in much of their lungs, which aids oxygen extraction. Crocodilian hearts are incipiently four-chambered but are still low pressure. Their lungs are internally dead end, but they, too, seem to have unidirectional airflow, and the method by which they are ventilated is sophisticated. Muscles attached to the pelvis pull on the liver, which spans the full height and breadth of the rib cage, to expand the lungs. This action is facilitated by an unusually smooth ceiling of the rib cage that allows the liver to glide back and forth easily, the presence of a rib-free lumbar region immediately ahead of the pelvis, and, at least in advanced crocodilians, a very unusual mobile pubis in the pelvis that enhances the action of the muscles attached to it.

Birds and mammals have fully developed, four-chambered, double-pump hearts able to propel blood in large volumes at high pressures. Mammals retain fairly large dead-end lungs, but they are internally very intricate, greatly expanding the gas-exchange surface area, and so are efficient despite the absence of one-way airflow. The lungs are operated by a combination of rib action and the vertical, muscular diaphragm. The presence of the diaphragm is indicated by the existence of a well-developed, rib-free lumbar region, preceded by a steeply plunging border to the rib cage on which the vertical diaphragm is stretched.

It is widely agreed that all dinosaurs probably had fully four-chambered, high-capacity, high-pressure hearts. Hearts make up to 0.5–1 percent of overall mass, so those of giant sauropods would have weighed hundreds of kilograms (twice as many pounds), comparable to whales of similar size. The super tall sauropods would have needed, like giraffes, special vascular adaptations—among them extra powerful muscles of the left ventricle that power arterial blood—to cope with the problems associated with fluctuating pressures from very high to neutral as the animal stood with the oxygen-demanding brain many meters above the heart and feet, or lay down, or raised and lowered its head from drinking level to the maximum vertical reach. During the latter, extra strength, tension-resistant tissues were needed to contain the high pressures below heart level, especially in the feet. Blood makes up about a twelfth of total body mass, so a 50 tonne sauropod would have contained about 4 tonnes, or 4,000 L (1,000 gal), of blood, again similar to great whales. Mammal red blood cells lack a nucleus, which increases their gas-carrying capability. The red blood cells of reptiles, crocodilians, and birds retain a nucleus, so those of these dinosaurs should have as well, leaving them a little inferior in O_2 capacity.

The respiratory complexes that oxygenated and decarbonized sauropodomorphs evolved considerably over time. Restoring the respiratory complexes of saurischians is aided by birds being members of the group. Birds have the most complex and efficient respiratory system of any vertebrate. Because the lungs are rather small, the chest ribs that encase them are fairly short, but the lungs are internally intricate, so they have a very large gas-exchange area. The lungs are also rather stiff and set deeply up into the strongly corrugated ceiling of the rib cage. The lungs do not dead-end; instead, they are connected to a large complex of air sacs whose flexibility and especially volume greatly exceed those of the lungs. Some of the air sacs invade the pneumatic vertebrae and other bones, but the largest sacs line the sides of the trunk; in most birds, the latter air sacs extend all the way back to the pelvis, but in some, especially flightless examples as observed in kiwis and ostriches, they are limited to the rib cage. The chest and abdominal sacs are operated in part by the ribs; the belly ribs tend to be extra long in birds that have well-developed abdominal air sacs. All the ribs are highly mobile because they attach to the trunk vertebrae via well-developed hinge articulations. The hinging is oriented so that the ribs swing outward as they swing backward, inflating the air sacs within the rib cage, and then deflating the sacs as they swing forward and inward. In most birds, the movement of the ribs is enhanced by ossified uncinate processes that form a series along the side of the rib cage. Each uncinate process acts as a lever for the muscles that operate the rib the process is attached to. In most birds, the big sternal plate also helps ventilate the air sacs. The sternum is attached to the ribs via ossified sternal ribs that allow the plate to act as a bellows on the ventral air sacs. In those birds with short sternums, the flightless ratites, and in active juveniles, the sternum is a less important part of the ventilation system.

The system is set up in such a manner that most of the fresh inhaled air does not pass through the gas-exchange portion of the lungs but instead goes first to the air sacs, from where it is injected through the entire lungs in one direction on its way out. Because this unidirectional airflow eliminates the stale air that remains in dead-end lungs at the end of each breath and allows the blood flow and airflow to work in opposite, countercurrent directions that maximize gas exchange, the system is very efficient. Some birds can sustain cruising flight at levels higher than Mount Everest and equaling those of jet airliners.

Neither the first theropods nor earliest prosauropods show clear evidence that they possessed air sacs, and aside from their lungs therefore being dead-end organs or close to it, little is known about their respiration. Theropods would go on to evolve increasingly extensive and capable systems until they reached the near-avian and then fully avian condition. Early euprosauropods, too, started to show signs of incipient air sacs. Sauropods show strong evidence that they independently developed an air-sac system approaching that of birds in complexity. The vertebrae were usually highly pneumatic, sometimes including the base of the tail. All the ribs were hinge-jointed, even the belly ribs, which one would expect to instead be solidly anchored in order to better support the big belly. A corrugated rib-cage ceiling favors rigid lungs. Most researchers agree that the air sac–filled vertebrae and mobile belly ribs of sauropods are strong signs that they had an air sac–driven respiratory complex that probably involved unidirectional airflow and approached, but did not fully match, the sophistication and efficiency of that possessed by birds. Lacking an elongated sternum and long aft ribs, the air

Long-necked sauropod and bird respiration

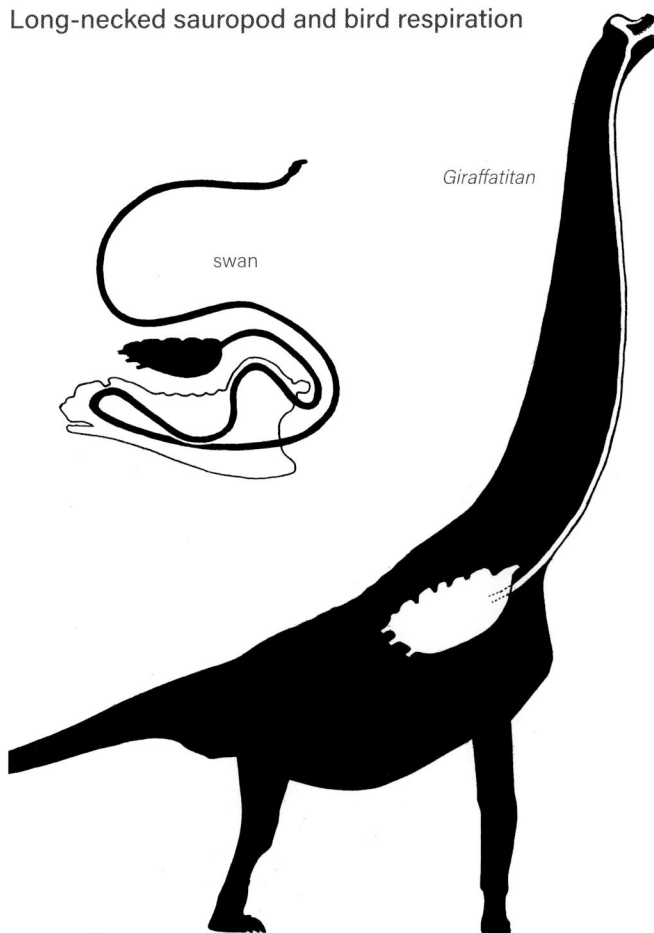

Giraffatitan

swan

sacs should have been limited to the rib cage, leaving the abdomen free to be entirely filled with digestive organs. Sauropods pose an interesting respiratory problem because most of them had to breathe through very long tracheae, which created a large respiratory dead space that had to be overcome with each breath. This is paralleled in long-necked birds, such as swans, which have a looped trachea in their chest, so an extra long trachea does not critically inhibit breathing. Presumably, the great air capacity of the air sacs helped sauropods completely flush the lungs with fresh air during each breath.

DIGESTIVE TRACTS

Herbivorous reptiles and birds lack the ability to thoroughly chew the plant materials they crop and ingest. Vegetation is swallowed with minimal oral processing and then broken down in the gut. Many plant-eating birds use gastroliths, aka gizzard stones, to help physically pulp the foliage by abrasion, crushing, and churning before further abdominal operations. In smaller birds, this is grit; in larger examples, such as ratites, gravel-sized hard stones, polished by constant tumbling and muscular contractions, are used in stone-rolling gizzard mills. Ornithischian dinosaurs paralleled herbivorous mammals in being oral processors that feature dental batteries covered by cheeks to extensively

Sauropod respiratory complex

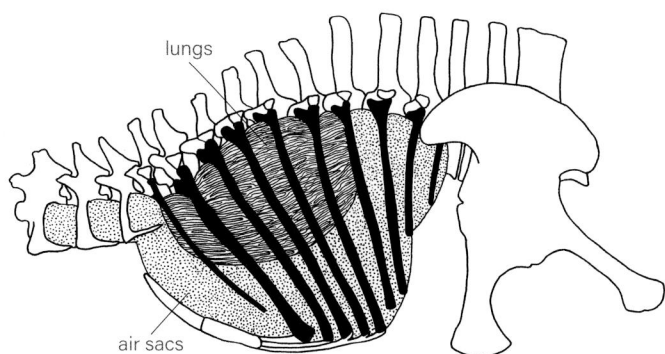

lungs

air sacs

Diplodocid and titanosaurid
sauropod front views

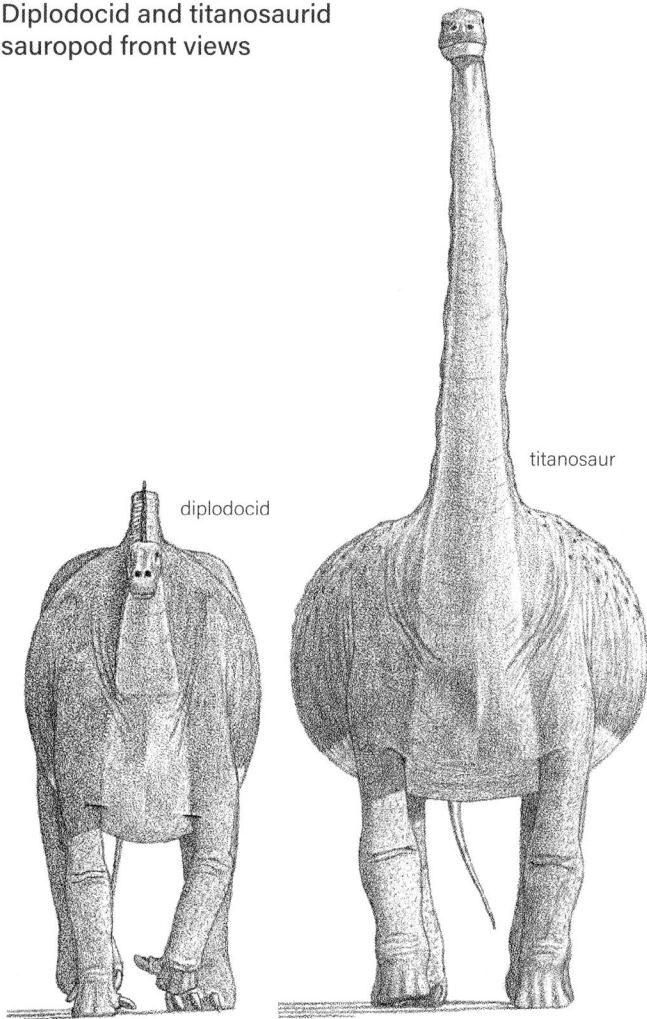

diplodocid

titanosaur

masticate plant stuffs before it is swallowed—or in the case of ruminants, after the material has undergone biochemical treatment in the gut and is regurgitated for physical breakdown by the teeth and reswallowed. The digestive organs make up in the area of a sixth of total mass in plant consumers. All herbivores use a bacterial gut flora to help deal with plant fibers and cellulous. The degree to which this is done depends on the nature of the digestive tract and the digestibility of the food—this is done at some cost, because the microbes use about a quarter of the nutrients and calories for themselves. The digestive complex of a gigantic sauropod would contain a tonne or so of symbiotic microbes. At a given size, tachyenergetic endotherms with high metabolic rates and aerobic budgets need to consume and digest more food than bradyenergetic endotherms, up to 10 times as much or more.

Lacking food-grinding dental batteries, sauropodomorphs were not sophisticated oral processors. The majority of prosauropods had narrow, pointed mouths as part of lightly constructed, moderately muscled skulls suitable for browsing leaves and twigs that were on the delicate side. A few prosauropods had stouter skulls able to take on somewhat tougher foliage. The cheeks that appear to have been present on at least some prosauropods and early sauropods should have allowed them to mash food before swallowing. Modest food chewing may explain why prosauropods did not have particularly big, fermenting tummies.

Sauropodomorph abdominal evolution was a story of increasingly plump bellies over evolutionary time. The first sauropods switched to more massive digestive complexes than prosauropods, with larger gut floras that could more extensively and efficiently digest tougher vegetation that was gathered in most cases by bigger, stronger teeth. The cheeks were lost. Tooth wear was often extensive, reflecting the cropping of rougher foliage, including thicker branches, as well as grit picked up with ground cover. Diplodocoids and titanosaurs converted to radically altered heads, with the teeth converted to slender pencils set tightly together at the front of a more squared-off mouth that allowed rapid cropping of foliage from both branches and the ground. The most straight-across jaws were largely for grazing, like broad-lipped white rhinos. The result was very fast tooth wear that was made up for with very rapid replacement of the teeth. Titanosaurs went the furthest with their enormous plant-processing bellies; approaching the end of the Mesozoic, all sauropods were broad-beamed. Well-rounded, hard stone, gastroliths found in association with sauropodomorphs and littering the sediments indicate they helped the food processing by churning the food in the gizzard. In healthy herbivores, the ingested fodder, bacterial gut flora, feces, and gizzard stones if present, make up 10–20 percent of the animal's total mass. There is no evidence that sauropods evolved an ultra-efficient ruminant-like cud chewing system like those of artiodactyl ungulates; such work only in animals of medium size in any case. As a side effect of being nonruminants, sauropodomorphs would not have released large quantities of methane into the atmosphere.

Late Jurassic sauropod
gastrolith, actual size

Predaceous
Eoraptor

A new finding of the first fossilized gut contents—a cololite—with a Cretaceous Australian specimen confirms the general thinking about sauropod diets. That the plant materials were not well ground up means they were not chewed much before swallowing. Conifer fodder likely came from the crowns of trees. Angiosperm leaves were probably from mid-level plants. And seed-fern fruiting bodies suggest cropping ground cover. This combination indicates this early titanosaur was a generalist. Various other sauropods should have been more specialized in their dietary habits.

Having some sharp teeth, the first prosauropods were omnivores. But the all blunt-toothed euprosauropods and sauropods could have readily reached out with their long necks and snapped up unwary protein- and calcium-rich creatures small enough to swallow whole from the ground and in vegetation, as do ratite birds. The victims could have included lesser dinosaurs and juveniles, including the odd baby sauropod.

Sauropodomorphs needed to drink copious quantities of water to supplement what they obtained via fresh vegetation. For a gigantic sauropod, this could be a few hundred liters of water a day (a fourth as much in gallons), most of all when it was hot, and much of the H_2O was used for cooling. Mammals can suck up water in quick order. Among diapsids, only pigeons are known to do so; the rest have to lift their heads and let the water drain down their gullets. Repeating this motion could have been awkward for sauropodomorphs with necks meters long, and spending time at the edge of water can be dangerous due to crocs and such. To solve these problems, it is plausible that the dinosaur group evolved a water-pump system.

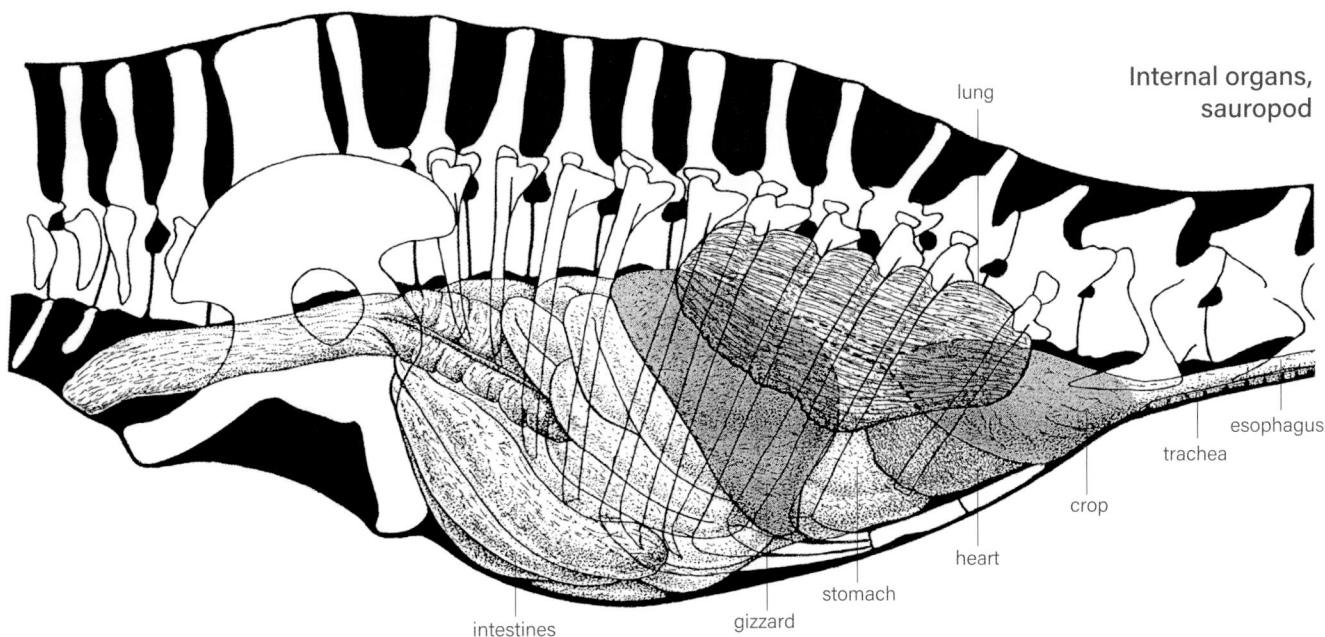

Internal organs, sauropod

lung

esophagus

trachea

crop

heart

stomach

gizzard

intestines

SENSES

The large eyes and well-developed optical lobes characteristic of most dinosaurs indicate that vision was usually their primary sensory system, as it is in all birds. Any binocular vision tied to the overlapping fields of vision in prosauropods and sauropods would have been modestly developed at best. Reptiles and birds have full color vision extending into the UV range, so dinosaurs very probably did, too. The comparatively poorly developed color vision of most mammals is a heritage of the nocturnal habits of early mammals, which reduced vision in the group to such a degree that eyesight is often not the most important of the senses. Our own skew toward color vision is different from that typical of animals, being more oriented to green. Reptile eyesight is about as good as that of well-sighted mammals, and birds tend to have very high-resolution vision, both because their eyes tend to be larger than those of reptiles and mammals of similar body size and because they have higher densities of light-detecting cones and rods than mammals. The cones and rods are also spread at a high density over a larger area of the retina than in mammals, in which high-density light cells are more concentrated at the fovea (so our sharp field of vision covers just a few degrees). Some birds have a secondary fovea. Day-loving raptors can see about three times better than people, and their sharp field of vision is much more extensive, so birds do not have to point their eye at an object as precisely as mammals to focus on it. Birds can also focus over larger ranges, 20 diopters compared to 13 diopters in young adult humans. The vision of the bigger-eyed dinosaurs, including large sauropods, may have rivaled this level of performance. The dinosaurs' big eyes have been cited as evidence for both daylight and nighttime habits. Large eyes are compatible with either lifestyle—it is the (in this case unknowable) structure of the retina and pupil that determines the type of light sensitivity. Sauropodomorph eyes faced to the sides, maximizing the area of visual coverage for detecting potential threats at the expense of the binocular view directly ahead, which the herbivores did not need.

Most birds have a poorly developed sense of smell, the result of the lack of utility of this sense for flying animals, as well the lack of space in heads whose snouts have been reduced to save weight. Exceptions are some vultures, which use smell to detect rotting carcasses hidden by deep vegetation, and grub-hunting kiwis. As nonfliers with large snouts, many reptiles and mammals have very well-developed olfaction, sometimes to the degree that it is a primary sensory system, canids being a well-known example. In the dinosaur mold, prosauropods and sauropodomorphs often had extremely well-developed, voluminous nasal passages, with abundant room at the back for large areas of olfactory tissues. In prosauropods and some sauropods, such as camarasaurs, the olfactory lobes of the brain were large, verifying their highly effective sense of smell. Those of diplodocoids indicate more average olfactory abilities.

Herbivorous dinosaurs probably had to be approached from downwind to keep them from sensing their attackers and going on alert.

Mammals have exceptional hearing, in part because of the presence of large, often-movable outer ear pinnae that help catch and direct sounds into the ear opening, and especially because of the intricate middle ear made up of three elements that evolved from what were once jaw bones. In some mammals, hearing is the most important sense, bats and cetaceans being the premier examples. Reptiles and birds lack fleshy outer ears, and there is only one inner ear bone. The combination of outer and complex inner ears means that mammals can pick up sounds at low volume. Birds partly compensate by having more auditory sensory cells per unit length of the cochlea, so sharpness of hearing and discrimination of frequencies are broadly similar in birds and mammals. Where mammalian hearing is markedly superior is in the detection of high-frequency sound. In many reptiles and birds, the auditory range is just 1–5 kHz; owls are exceptional in being able to pick up from 250 Hz to 12 kHz, and geckos go as high as 10 kHz. In comparison, humans can hear 20 kHz, dogs up to 60 kHz, and bats 100 kHz. At the other end of the sound spectrum, some birds can detect very low frequencies: 25 Hz in cassowaries, which use this ability to communicate over long distances, and just 2 Hz in pigeons, which may help them detect approaching storms. It has been suggested that cassowaries use their big, pneumatic head crests to detect low-frequency sounds, but pigeons register even basser sounds without a large organ.

In the absence of fleshy outer and complex inner ears, dinosaur hearing was in the reptilian-avian class, and sauropodomorphs could not detect very high frequencies. Nor were the auditory lobes of these dinosaurs' brains especially enlarged, although they were not poorly developed, either. Nocturnal, flying, rodent-hunting owls are the only birds that can hear fairly high-frequency sounds, so certainly most and possibly all dinosaurs could not hear them, either. The big ears of sauropods had the potential to capture very low frequencies, allowing them to communicate over long distances, as can their elephant analogs. It is unlikely that hearing was the most important sense in any dinosaur, but it was probably important for detection of prey and predators, and for communication, in all species.

VOCALIZATION

No living reptile has truly sophisticated vocal abilities, which are best developed in crocodilians. Some mammals do, humans most of all. A number of birds have limited vocal performance, but many have evolved a varied and often very sophisticated vocal repertoire not seen among other vertebrates outside of people. Songbirds sing, and a number of birds are excellent mimics, to the point that some can imitate artificial sounds, such as bells and sirens, and parrots can produce understandable human-like speech. Some birds, swans particularly, possess elongated tracheal loops in the chest that they use to produce high-volume vocalizations. Cassowaries call one another over long ranges with very low-frequency sounds, and so do elephants. Birds possess the intricate voice boxes needed to generate complex vocalizations. Among dinosaur fossils, only an ankylosaur skull includes a complete voice box. The complicated structure of the armored dinosaurs' larynx suggests vocal performance at an avian level, perhaps high-end performance, and such may have been true of other dinosaurs. The long trachea of prosauropods and especially the long-necked sauropods should have been able to generate powerful low-frequency sounds that could be broadcast over long ranges. Vocalization is done through the open mouth, rather than through the nasal passages, so complex nasal passages acted as supplementary resonating chambers. It is doubtful that any nonavian dinosaur had vocal abilities to match the more sophisticated examples seen in the most vocally sophisticated birds and mammals. Although we will never know what dinosaurs sounded like, and the grand roars of dinosaur movies are not likely, there is little doubt that most Mesozoic forests, prairies, and deserts were filled with the sounds of sauropodomorphs.

GENETICS

As more fossils are found in different levels of geological formations, the evidence is growing that dinosaurs, sauropodomorphs included, enjoyed high rates of speciation that boosted their diversity at any given time. And over time, via a rapid turnover of species, most did not last for more than a few hundred thousand years before being replaced by new species one way or another. The same is true of birds, which have more chromosomes than slower-evolving mammals. Prosauropods and sauropods had the same genetic diversity as their direct avian descendants, which may have been a driving force behind their multiplicity.

DISEASE AND PATHOLOGIES

Planet Earth was infested with a toxic soup of diseases and other dangers that put all dinosaurs at high risk. The disease problem was accentuated by the global greenhouse effect, which maximized the tropical conditions that favored disease organisms, especially bacteria and parasites. Biting insects able to spread assorted diseases were abundant during the Mesozoic; fossils have been found in amber and fine-grained sediments. Reptile and bird immune systems operate somewhat differently from those of mammals; in birds, the lymphatic system is particularly important. Presumably, the same was true of their dinosaur relations.

Some skeletal pathologies appear to record internal diseases and disorders. Fused vertebrae are fairly common. Among them are a cofused pair of camarasaur neck-base bones that apparently froze together late in life, perhaps ossification of arthritis; being fixed upturned may have made it difficult for the animal to bend its neck down far enough to drink. The pneumatic space in a sauropod neck shows sign of a common avian respiratory fungal infection. Also found are growths that represent benign conditions or cancers. Most pathologies are injuries caused by stress or wounds; the latter often became infected, creating long-term, pus-producing lesions that affected the structure of the bone. Injuries tell us a lot about the activities of dinosaurs. Healed bite marks in the tails of sauropods indicate that they survived attacks by allosaurs and tyrannosaurs. A prosauropod apparently lost the last third of its tail to an attack that it survived at least initially. Some sauropods' ribs were fractured. Some sauropod foot bones show signs of stress fractures. Over all, despite or perhaps because of sauropods' size and slow speeds, they show relatively little evidence of impact injury. That said, some of the dinosaurs presumably died either slowly or quickly from the pathologies we find in their fossil remains.

BEHAVIOR

BRAINS, NERVES, AND INTELLIGENCE

Assessing brain power is complicated because many factors are involved. One that has long been used is the mass of the brain relative to total body mass at a given size. Within the context that brains of a given performance level tend to become smaller relative to the body as size increases—elephant brains are many times absolutely larger than those of people while being many times smaller relative to body weight, and we are over all more intelligent—relatively bigger brains are likely to produce higher cognition. Also important is brain structure, with birds and mammals having more complex schemes, including large forebrains. Adding to the complications is the neural density factor. Reptiles have much lower neural density relative to brain mass than mammals and birds, and the latter are markedly higher in this regard than mammals. The last point helps explain why birds with absolutely small brains, such as crows and parrots, achieve levels of thinking comparable to those of some far larger-headed primates. Avian brains are also markedly more energy efficient, their neurons requiring less glucose to process information. Big brains packed with lots of neurons can correlate with metabolism in that low-energy animals cannot produce enough metabolic power to operate high-cognition brains, which require a high metabolism. Less clear is whether energetic animals automatically have similarly energetic brains. In particular, it is not known whether reptilian brains can have high neural densities even if the animals run at high metabolic rates.

The brains of prosauropods and sauropods were reptilian both in size relative to the body and in structure. Even small-brained animals can achieve remarkable levels of mental ability. Fish and lizards can retain new information and learn new tasks. Many fish live in organized groups. Crocodilians care for their nests and young. Social insects with tiny neural systems live in organized colonies that rear the young, enslave other insects, and even build large, complex, architectural structures. It is not unthinkable that dinosaurs up to the biggest sauropods could use sticks and leafy branches to scratch themselves if they could reach close enough to their bodies with their mouths, use heavy sticks to knock down otherwise-unreachable choice food items, or build leafy branch piles over water holes to protect them when not in use, as elephants do.

The enlarged spinal cavity in the pelvic region of many small-brained dinosaurs was an adaptation to coordinate the function of the hindlimbs and is paralleled in big ground birds. The great length of some dinosaurs posed a potential problem in terms of the time it took for electrochemical impulses to travel along the nerves. In the biggest sauropods, a command to the end of the tail and the response back could have to travel as much as 75 m (250 ft) or more. Synaptic gaps where chemical reactions transmit information slow down the impulses, so this problem could have been minimized by growing individual nerve cords as long as possible.

SOCIAL ACTIVITIES

Land reptiles do not form organized groups. Birds and mammals often do, but many do not. Most big cats, for instance, are solitary, but lions are highly social. Some, but not all, deer form herds.

That sauropodomorphs often formed social groups is supported by some single-species bone beds that do not appear to have been death traps and slowly accrued fossils over time, or perhaps resulted from droughts that compelled numerous individuals to gather at a water source where they starved to death as the vegetation ran out. Some accumulations appear to have been the result of sudden events caused by volcanic ashfalls, flash floods, drownings when many dinosaurs crossed fast-flowing streams, or dune slides. Such bone beds, which in some cases suggest the existence of very large herds, usually consist of herbivorous dinosaurs.

Trackways are the closest thing we have to motion pictures of the behavior of fossil animals. A significant portion of the trackways of prosauropods and sauropods show the prints of solitary animals, indicating that the makers were not part of a larger group. A number of trackways lie close together on parallel paths. In some cases, this may be because the track makers were forced to follow the same path along a shoreline even if they were moving independently of one another. But sometimes the parallel trackways are crisscrossed by the trackways of other dinosaurs that appear to have been free to travel in other directions. Some sauropod prints were laid in tight groups that are the harbingers of animals deliberately moving as a joint collective containing up to dozens of individuals.

The degree of organizational sophistication of the groups was not as high as those of birds and mammals, which normally contain the offspring of the parents. Sauropod herds were probably more similar to fish schools. A key clue is that the great herd formations lack small juveniles—unable to keep up with the enormous grown-ups and subject to being trampled upon, juveniles under a tonne moved in their own pods. Suggestions that the trackways of sauropods show that the juveniles were ringed by protective parents have not been borne out, and it is unlikely that very large dinosaurs directly cared for and protected offspring that were so tiny when they came out of their eggs or nests.

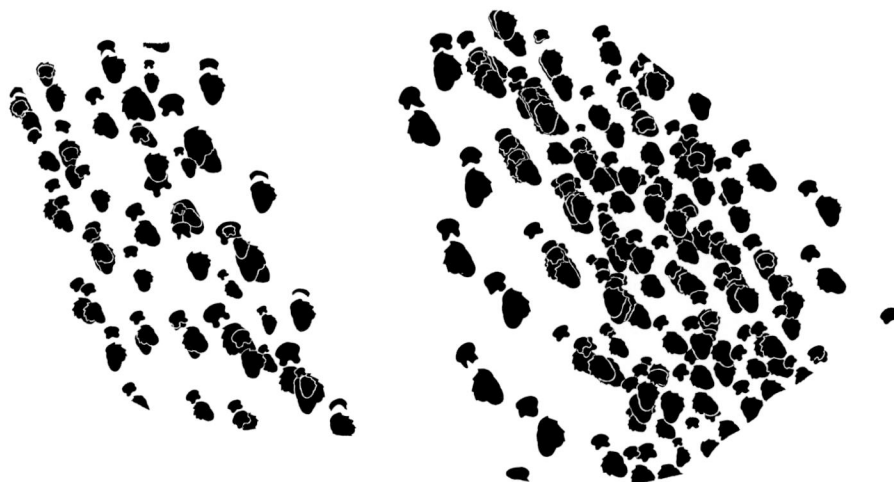

Trackways of
a herd of sauropods

REPRODUCTION

It has been suggested that some dinosaur species exhibit robust and gracile morphs that represent the two sexes. It is difficult either to confirm or to deny many of these claims because it is possible that the two forms represent different species. There is as yet little if any evidence that dimorphism was present into sauropodomorphs, although this can by no means be ruled out.

Reptiles and some birds and mammals, including humans, achieve sexual maturity before reaching adult size, but most mammals and extant birds do not. Sauropodomorphs followed the reptilian scheme—for giant sauropods to wait decades until they reached full size before replacing themselves would have been maladaptive.

The sauropodomorphs lacked many of the specialized display devices in terms of bone-based horns, head crests and frills, and the like found on many dinosaurs. What they did have were their usually long necks and tails, which could be used for display, midline frills, perhaps wattles in some examples, and sheer size in the case of big examples. Some or all prosauropods and sauropods could have engaged in intricate ritual display movements and vocalizations during competition and courtship that have been lost to time. Because height as reflected by the neck was probably a prime form of display—including via rearing—displays may have been prone to be frontal. The ability of titanosaurs to use the ball-and-socket tail-base vertebrae to curl their whip-tipped tails over their backs may have been a display device for mating purposes. While the brachiosaurs, with their high shoulders, towering necks, and small tails, certainly would have put emphasis on frontal, the diplodocoids are harder to parse out. Their large whip tails look like fine display organs, but diplodocoids may have emphasized rearing to make an impression, in which case the tail would be on the ground—possibly, they had complicated display routines in which necks were emphasized tripodally and then tails in lateral presentations. If diplodocoid and titanosaur tail whips could go supersonic to generate sonic cracks, as some propose and others oppose, the sounds could have been an auditory supplement to intraspecific presentations. Very likely, sauropods used their booming voices to make an impression on opponents and potential mates. While intraspecific competition is often peaceful in order to minimize damage to the participants, it can be forceful and even violent, as per battling male hippos that may kill one another. Rearing sauropodomorphs could have assaulted one another with their thumb claws if they had them; tail-bouncing prosauropods could also lash out with their hindclaws. Those sauropods with small tail clubs probably wacked each other's flanks.

In reptiles and birds, the penis or paired penises (if either are present) and the testes are internal, and this was the condition in dinosaurs. Most birds lack a penis, but whether any dinosaur shared this characteristic is unknown. Presumably, copulation was a quick process that occurred with the female lowering her shoulders and swinging her tail aside to provide clearance for the male, which reared behind her on two legs or even one leg while placing his hands on her back to steady himself. The need of sauropods to copulate supports the ability of these giants, including the biggest and those with long forelimbs, to stand on the hindlegs alone.

At least some dinosaurs, including sauropods, produced hard-shelled eggs like those of birds, rather than the softer-shelled eggs of reptiles, including crocodilians, and monotremes. The evolution of calcified shells may have precluded live birth, which is fairly common among reptiles and is absent in birds. On the other hand, eggs of prosauropods appear to have been soft-shelled, indicating that there was considerable variation in the feature in dinosaurs, perhaps even within and between subgroups. If so, that could help explain why remains of dinosaur eggs are surprisingly scarce through much of the Mesozoic. For example, not a single eggshell fragment attributable to the many sauropod species that inhabited the enormous Morrison Formation has yet been found. Firmly identifying the producer of a given type of egg requires the presence of intact eggs within the articulated trunk skeleton, or identifiable embryo skeletons within the eggs. Prosauropod and sauropod eggs were subcircular. Those of the former were about 60–70 mm in diameter (2–3 in) and 100 g (3.5 oz), about twice the size of chicken eggs and broadly similar to crocodilians and big monitor lizards. The only sauropod eggs are those of titanosaurs—what other sauropods were doing remains oddly unknown. They ranged from 120 to 300 mm (up to a foot) and from 0.6 to 10 kg (up to 22 lb). The largest known eggs are those of the elephant bird *Aepyornis maximus*, which were a little larger than the observed sauropod maximum—the biggest eggs of a living bird are those of ostriches, at 150 mm (half a foot) and 0.7 kg (1.5 lb). Incubation periods for large reptile eggs range from a couple to many months, and such would have been true for prosauropods and sauropods, a half or year or substantially more being plausible for their bigger eggs. Some ratites create communal nests in which more than one female lays their eggs, and this may have been true of some dinosaurs.

There are two basic reproductive stratagems: r-strategy and K-strategy. K-strategists are slow breeders that produce few young; r-strategists produce large numbers of offspring that offset high losses of juveniles. Rapid reproduction has an advantage: producing large numbers of young allows a species to expand its population quickly when conditions are suitable, so r-strategists are "weed species" able to rapidly colonize new territories or promptly recover their population after it has crashed for one reason or another. Sauropodomorphs were r-strategists that typically laid large numbers of eggs in the breeding season. This helps explain why breeding sauropodomorphs laid much smaller eggs than birds at a given adult size. With supersauropods putting out eggs not as large as those of superbirds a hundred times lighter, the sauropodomorphs were emphasizing egg numbers over size in the reptilian mode. By contrast, birds produce a modest number of relatively large eggs and provide the chicks with considerable parental attention. One r-strategist bird group is the big modern ratites, which produce numerous eggs to overcome the high predation rate of their offspring while inside and later outside the eggs. This is in contrast to the great island ratites, which laid only one to a few oversized eggs a year because the young were not at risk of being snarfed up by predators until humans liquidated the populations in part by eating their delicious supereggs. Sauropods appear to have placed the largest number of eggs in a single nest, up to a few dozen. The physiological demands of laying down so much calcium so quickly in the forming shells may be a reason titanosaurs had armor osteoderms to tap into for extra calcium during egg formation. The fast-breeding dinosaurs were very different from giant mammals, which are K-strategists that produce few calves that then receive extensive care over a span of years, including nursing the young via milk-producing mammary glands.

The hatchlings of all reptiles are precocial, having bones and joints ossified well enough, and muscles strong enough, to be able to leave the nest immediately after getting out of their eggs. Baby sea turtles demonstrate this when they immediately skitter across the beach into the waves. Birds are much more variable. Some are very precocial from the get-go; megapode fowl chicks can fly shortly after hatching. Others are highly altricial, unable to leave the nest for extended periods and dependent upon their parents to bring them food until they fledge.

It was long tacitly assumed that, like most reptiles, dinosaurs paid little or no attention to their eggs after burying them. A few lizards do stay with the nest, and pythons actually incubate their eggs with muscle heat. Crocodilians often guard their nests and hatchlings. All birds lavish attention on their eggs. Nearly all incubate the eggs with body heat; the exception is megapode fowl, which warm eggs in mounds that generate heat via fermenting vegetation. The fowl carefully regulate the temperature of the nest by adding and removing vegetation to and from the mound. But when megapode chicks hatch, they are so well developed that the precocial juveniles quickly take off and survive on their own.

Reptiles lay their large numbers of eggs in pairs, and sauropodomorphs did so as well, not singly as per birds. Prosauropod and sauropod eggs were buried in rather irregular nests in the reptilian norm. Hand and foot claws helped in digging the shallow, broad pits. In some prosauropod nests, many eggs are tightly packed in rows. It is difficult to see what purpose this would have served, so it is not certain if the mother deliberately arranged them that way. With the big-bellied parents too large to brood their eggs, incubation was via ground heat, possibly facilitated by fermenting vegetation included in mulch spread atop the nest. Nests were in colonies, presumably because the soil conditions in the location were optimal, and putting so many nests in the same place would have overwhelmed the local egg eaters. There is evidence that at least some sauropods deposited their eggs near geothermal heat sources.

The known skeletons of prosauropod hatchlings have poorly ossified bones and joints, indicating they were altricial babies unable to leave their nests immediately. This in turn indicates that their parents stayed near the nests, providing protection against egg stealers and possibly monitoring the nest's temperature and adjusting the amount of material above the eggs to keep them at the optimal heat level, like megapode fowl. Once hatched, the nestlings should have been dependent upon their parents for food for weeks up to a few months. Not being especially large, the nesting parents would not have stripped the local flora bare, and the amount of forage needed by nestlings of a few kilograms or less would have been trivial—unlike small birds, which have to work frantically all the daylight hours to keep their perpetually ravenous altricial nestlings satisfied. Because foraging required leaving the nest, it is probable that more than one parent was involved in caring for their charges, one of them staying at the nest to protect the contents while the other was away. The later duck-billed hadrosaurs appear to have practiced a similar form of parental nestling care. It is plausible that at least some of the smaller sauropods did so, too.

Giant adult sauropods would have risked denuding the local vegetation as well as squashing their own eggs if they remained to guard their nests. Also in danger of being trampled were the hatchlings, which at only a few kilograms were thousands of times less massive than their parents. Laying so many eggs in so many nests helped overwhelm the ability of the local predators to find and eat all the eggs and emerging hatchlings, although a fossil shows a large snake feeding on a just-emerged sauropod hatchling. Some prehatchling reptiles in mass nests start vocalizing to coordinate their synchronous emergence, even though doing so risks attracting egg and hatchling eaters. Sauropod parents leaving their nests to their fates explains why trackways show the small, precocial juveniles formed their own pods. The inevitable high losses of these slowly ambling, weakly armed youngsters were made up for with sheer numbers. Wandering sauropod babies may have been at risk of being picked off at random by grown members of their own species. The young of giant sauropods joined full-sized adults only after a few years, when they had reached hundreds of kilograms. The mature sauropods probably paid the young ones no particular notice and were unlikely to have even been closely related to them. In this scenario, the juveniles were seeking the statistical safety of being near aggressive grown-ups able to battle the big predators. In the best-known herd trackway, there are only a few really big grown-ups—one of 50 tonnes, another of 30 tonnes, five 20 tonners, the rest on down to over a tonne. This is entirely different from the much more sophisticated herds of elephants, which are made up mainly of adult females collectively raising their

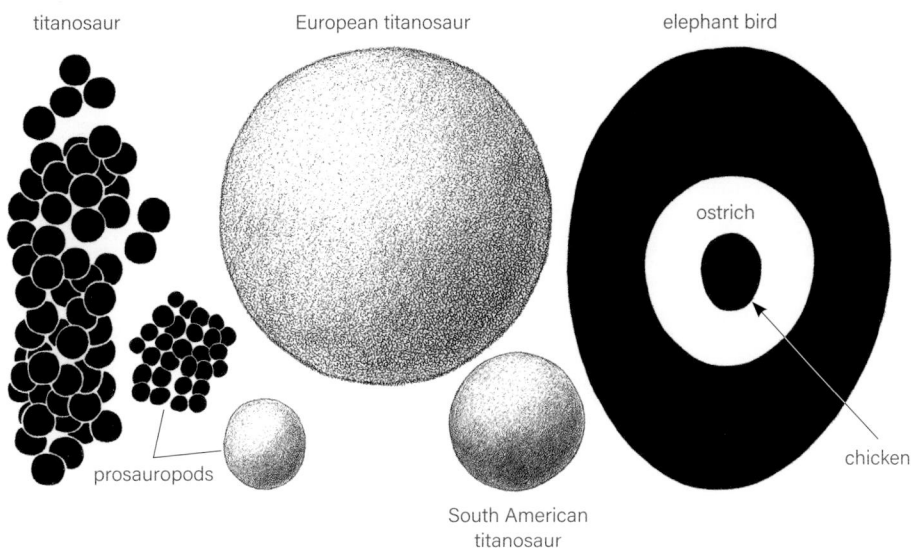

Eggs and nests

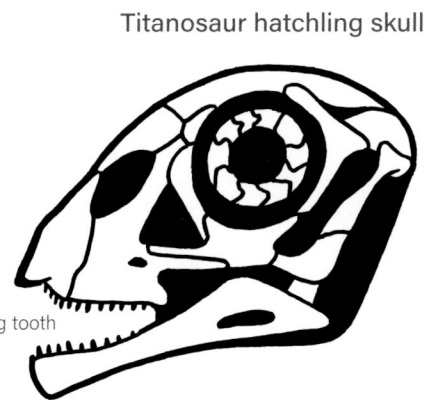

titanosaur

European titanosaur

elephant bird

prosauropods

South American
titanosaur

ostrich

chicken

Titanosaur hatchling skull

egg tooth

prosauropod

Hatchlings

titanosaur

their own young but also young-sters that were not their own.

Because no sauropodomorphs, even the prosauropods, lavished off-spring with anything close to the parenting typical of lactating mam-mals, their progeny could grow up on their own even if the adult popula-tion crashed. A juvenile mammal is likely dead if it loses its mother, as is true of many bird chicks.

A problem that all embryos that develop in hard-shelled eggs face is getting out of that shell when the time is right. The effort to do so is all the harder when the egg is large and the shell correspondingly thick. Fortunately, some of the shell is absorbed and used to help build the skeleton of the growing creature. Baby birds use an egg tooth to achieve the breakout. The same has been found adorning the nose of titanosaur sauropod embryos, as was the likely norm for the entire group.

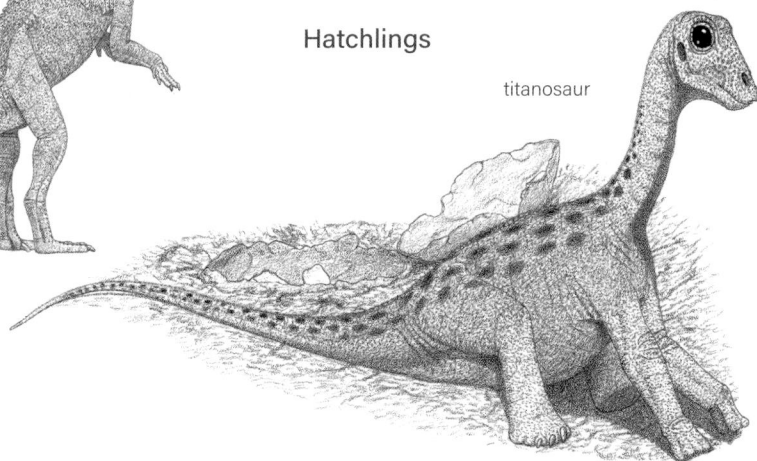

few calves, including newborns. Prosauropod parents may have been able to allow their offspring to tag along in fa-milial groups once they were large enough to leave the nest, similar to ratites and some other birds. As the juveniles fed themselves, they would have enjoyed the intentional pro-tection of their parents. If any prosauropods nested com-munally, then the adults would be looking after not only

GROWTH

All land reptiles grow slowly. This is true even of giant tor-toises and big, energetic (by reptilian standards) monitors. Land reptiles can grow most quickly only in perpetually hot equatorial climates, and even then they are hard-pressed

to reach a tonne. Aquatic reptiles can grow more rapidly, probably because the low energy cost of swimming allows them the freedom to acquire the large amounts of food needed to put on bulk. But even crocodilians, including the extinct giants, which reached nearly 10 tonnes, do not grow as fast as many land mammals. Mature reptiles tend to continue to grow slowly throughout their lives.

Some marsupials and large primates, including humans, grow no faster or only a little faster than the fastest-growing land reptiles. Other mammals, including other marsupials and a number of placentals, grow at a modest pace. Still others grow very rapidly; horses are fully grown in less than two years, and aquatic whales can reach 50–100 tonnes in just a few decades. Bull elephants take about 30 years to mature. All living birds grow rapidly; this is especially true of altricial species and of the big ratites. No extant bird takes more than a year to grow up, but some of the recently extinct giant island ratites may have taken a few years to complete growth. The secret to fast growth appears to be having an aerobic capacity high enough to allow the growing juvenile, or its adult food provider, to gather the large amounts of food needed to sustain rapid growth.

High mortality rates from predation, disease, and accidents make it statistically improbable that unarmored, nonaquatic animals will live very long lives, so they are under pressure to grow rapidly. On the other hand, starting to reproduce while still growing tends to slow down the growth process as energy and nutritional resources are diverted to produce offspring. Few mammals and no living birds begin to breed before they reach adult size. No bird continues to grow once it is mature. Nor do most mammals, but some marsupials and elephants never quite cease growing.

At the microscopic scale, the bone matrix is influenced by the speed of growth, and the bone matrix of sauropodomorphs tended to be more similar to that of birds and mammals, which grows at a faster pace than that of reptiles. Bone ring counts are being used to estimate the growth rate and life span of the dinosaurs, but this technique can be problematic because some living birds lay down more than one ring in a year, so ring counts can overestimate age and understate growth rate. There are additional statistical issues, because as animals grow, the innermost growth rings tend to be destroyed, leading to

Diplodocus growth series

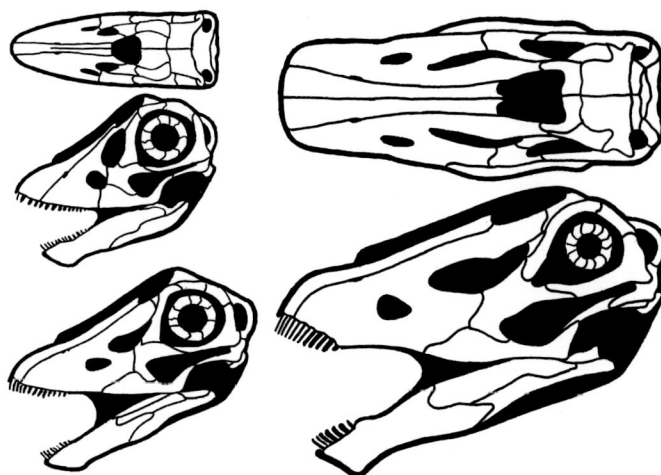

difficulties in estimating the number of missing age markers. The sauropodomorphs sampled so far appear to have grown at least somewhat faster than land reptiles. Rates of growth were variable in prosauropods, as was final adult size to a remarkable degree. Modestly sized sauropods appear to have grown as fast as similarly sized land mammals. Some giant sauropods appear to have grown with the spectacular swiftness seen in the big rorqual whales, getting to full size in a few decades—note that a *Jurassic Park* scenario flaw as little noticed as it is glaring is the presence of giant artificially bred dinosaurs so soon after the initiation of the paleozoo project. Titanosaurs may have been the fastest to grow up. The growth achievement of the greatest sauropods is astonishing. Giant mammals get a head start, being born as large calves only a few dozen times smaller than the adults, and then being nourished with enormous amounts of nutrient-dense milk. Hatchling sauropods had to expand their mass tens of thousands of times in

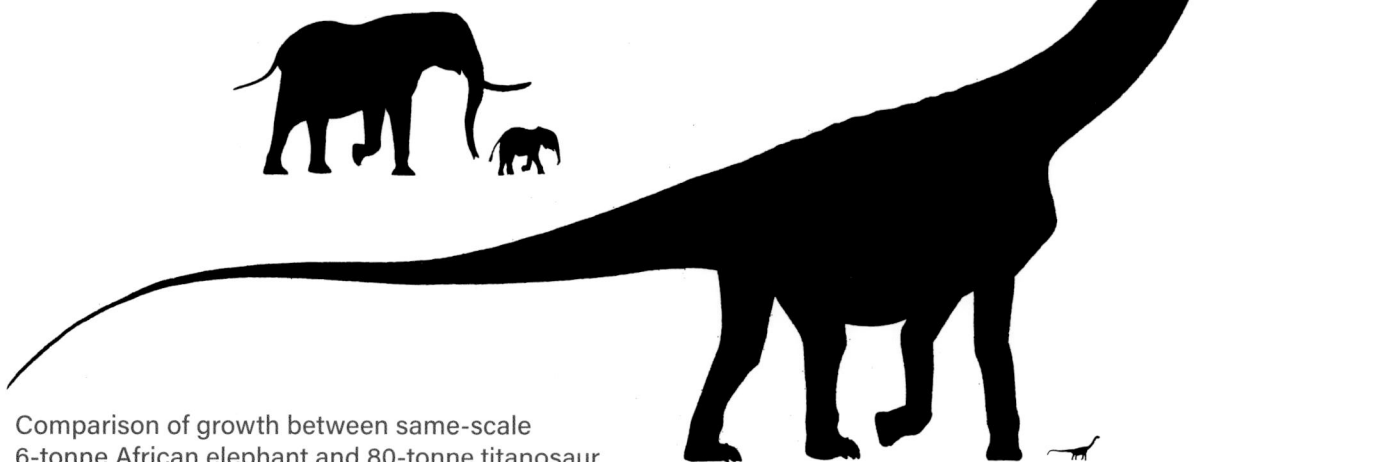

Comparison of growth between same-scale 6-tonne African elephant and 80-tonne titanosaur

just a few decades and with little or no nourishment provided by the adults.

Although they do not undergo metamorphosis like many invertebrates, fish, and amphibians, many amniote tetrapods experience considerable allometric modification in body form and proportions as they grow up, humans being an example. Relative head size is typically larger at the beginning, and it is common for large land animals to start out as lightly constructed and gracilely legged and become more robust in body and especially limbs with maturity—gangly foals compared to adult horses are a paragon of that pattern. Sauropods are notable in that they did not change all that much with growth, being fairly isometric in body, limb, and tail proportions. The load-bearing femur was not particularly slender in even small juveniles and was quite attenuated in some gigantic examples. This exceptionally low degree of change—even in view of the extreme change in total size in many, albeit not all, sauropods—reflects their slow gait at all sizes, like elephants whose proportions remain fairly similar as they age. Sauropod heads did become somewhat smaller with greater size, and necks longer, but even the teenagers had small heads and their necks were never short. Teen to adult camarasaur heads were isometric, because even as adults their snouts were short, the retention of a juvenile feature into adulthood being neoteny. Growing diplodocoids and titanosaurs underwent substantial snout elongation and broadening as their heads became increasingly horse-shaped; the juveniles' relatively narrow snouts reflected a shift from being selective feeders of digestible plant items to courser materials. While growing prosauropods did not multiply their size as much as sauropods, they underwent more extensive alterations more typical of amniotes. Their heads started out relatively large, with a very short rostrum and short neck, and their arms became relatively shorter as adults became more bipedal.

The cessation of significant growth of the outer surface of many adult dinosaur bones indicates that most but not all species did not grow throughout life. Medium-sized and large mammals and birds live for only a few years or decades, elephants live about half a century, and giant whales can last longer, with the sluggish rights making it well over 100 years. There is no evidence sauropodomorphs lived longer than similarly tachyenergetic mammals or birds of the same size. Sauropods seem to have enjoyed similarly long life spans, although whether they made it to a full century or beyond is not documented—but would not be surprising in the case of the biggest examples. If they did, they may have matched the life spans of much smaller and less energetic giant tortoises.

ENERGETICS

Vertebrates can utilize two forms of power production. One is aerobiosis, the direct use of oxygen taken in from the lungs to power muscles and other functions. Like air-breathing engines, this system has the advantage of producing power indefinitely but is limited in its maximum power output. An animal that is walking at a modest speed for a long distance, for instance, is exercising aerobically. The other form of power production is anaerobiosis, in which chemical reactions that do not immediately require oxygen are used to power muscles. Rather like rockets that do not need to take in air, this system has the advantage of being able to generate about 10 times more power per unit of tissue and time. But it cannot be sustained for an extended period and produces toxins that can lead to serious illness if sustained at too high a rate for too long, which is tens of minutes. Anaerobiosis also builds up an oxygen debt that has to be paid back during a period of recovery. Any fairly fast animal that is running, swimming, or power flying near its top speed is exercising anaerobically.

Most fish and all amphibians and reptiles have low resting bradymetabolic rates and low aerobic capacity. They are therefore bradyenergetic, and even the most energetic reptiles, including the most aerobically capable monitor lizards, are unable to sustain truly high levels of activity for extended periods. Many bradyenergetic animals are, however, able to achieve very high levels of anaerobic burst activity, such as when a monitor lizard or crocodilian suddenly dashes toward and captures prey. Because bradyenergetic animals do not have high metabolic rates, they are largely dependent on external heat sources, primarily the ambient temperature and the sun, for their body heat, so they are ectothermic. As a consequence, bradyenergetic animals tend to experience large fluctuations in body temperature, rendering them heterothermic. The temperature at which reptiles normally operate varies widely depending on their normal habitat. Some are adapted to function optimally at modest temperatures of 12°C (54°F). Those living in hot climates are optimized to function at temperatures of 38°C (100°F) or higher, so it is incorrect to generalize reptiles as "cold-blooded." In general, the higher the body temperature, the more active an animal can be, but even warm reptiles have limited sustained activity potential.

Most mammals and birds have high resting tachymetabolic rates and high aerobic capacity. They are therefore tachyenergetic and are able to sustain high levels of activity for extended periods. The ability to exploit oxygen for power over time is probably the chief advantage of being tachyenergetic. Tachyenergetic animals also use anaerobic power briefly to achieve the highest levels of athletic performance, but they do not need to rely on this as much as reptiles, are not at risk of serious self-injury, and can recover more quickly. Because tachyenergetic animals have high metabolic rates, they produce most of their body heat internally, so they are endothermic. As a consequence, tachyenergetic animals can achieve more stable body temperatures. Some, like humans, are fully homeothermic, maintaining a nearly constant body temperature at all times when healthy. Many birds and mammals, however, allow their body temperatures to fluctuate to varying degrees, for reasons ranging from going into some degree of torpor to storing excess heat on hot days, on a daily or seasonal basis. So they are semi-heterothermic or semihomeothermic, depending on the degree of temperature variation. The ability to keep the body at or near its optimal temperature is another advantage of having a high metabolic rate. Normal body temperatures range from 30°C to 44°C (86°F–111°F), with birds always at least at 38°C (100°F). High levels of energy production are also necessary to do the cardiac work that creates the high blood pressures needed to be a tall animal.

Typically, mammals and birds have resting metabolic rates and aerobic capacities about 10 times higher than those of reptiles, and differences in energy budgets are even higher. However, there is substantial variation from these norms in tachyenergetic animals. Some mammals, among them monotremes, some marsupials, hedgehogs, armadillos, sloths, and manatees, have modest levels of energy consumption and aerobic performance, in some cases not much higher than those seen in the most energetic reptiles. In general, marsupials are somewhat less energetic than their placental counterparts, so kangaroos are about a third more energy efficient than deer. Among birds, the big ratites are about as energy efficient as similarly sized marsupials. At the other extreme, some small birds share with similarly tiny mammals extremely high levels of oxygen consumption even when their small body size is taken into account. On the big side of the spectrum, elephants and whales have metabolic rates that are a continuation of the tachymetabolic norm and are well above those expected in reptiles of the similar bulk. It is possible for very large bradymetabolic animals to retain most of the heat they generate with their bulk, so they can be low-metabolic-rate endotherms. The common practice to refer to high-energy animals simply as endotherms is correspondingly simplistic, so creatures with high resting and active metabolisms are correctly referred to as tachyenergetic.

Widely different energy systems have evolved because they permit a given species to succeed in its particular habitat and lifestyle. Reptiles enjoy the advantage of being energy efficient, allowing them to survive and thrive on limited resources. Tachyenergetic animals are able to sustain much higher levels of activity that can be used to acquire even more energy, which can then be dedicated to the key factor in evolutionary success, reproduction. Tachyenergy has allowed mammals and birds to become the dominant large land animals from the tropics to the poles. But reptiles remain very numerous and successful in the tropics and, to a lesser extent, in the temperate zones.

As diverse as the energy systems of vertebrates are, there appear to be things that they cannot do. All insects have low, reptilelike resting metabolic rates. When flying, larger insects use oxygen at very high rates similar to those of birds and bats. Insects can therefore achieve extremely high maximal/minimal metabolic ratios, allowing them to be both energy efficient and aerobically capable. Insects can do this because they have a dispersed system of tracheae that oxygenate their muscles. No vertebrate has both a very high aerobic capacity and a very low resting metabolism, because the centralized respiratory-circulatory system requires that the internal organs work hard even when resting in tachyenergetic vertebrates. An insect-like metabolic arrangement should not, therefore, be applied to dinosaurs. However, it is unlikely that all the energy systems that have evolved in land vertebrates have survived until today, so the possibility that some or all dinosaurs were energetically exotic needs to be considered.

The general assumption until the 1960s was that dinosaur energetics was largely reptilian, but most researchers now agree that dinosaurs' power production and thermoregulation were closer to those of birds and mammals. It is also widely agreed that because dinosaurs were such a large group of diverse forms, there was considerable variation in their energetics, as there is in birds and especially mammals.

Reptiles' nonerect, sprawling legs are suitable for the slow walking speeds of 1–2 km/h (0.5–1 mph) that their low aerobic capacity can power over extended periods. Sprawling limbs also allow reptiles to easily drop onto their belly and rest if they become exhausted. No living bradyenergetic animal has erect legs. Walking is always energy expensive—it is up to a dozen times more costly than swimming the same distance—so only aerobically capable animals can easily walk faster than 3 km/h (2 mph). The long, erect legs of sauropodomorphs matched those of birds and mammals and favored the high walking speeds of 3–10 km/h (2–6 mph) that only tachyenergetic animals can sustain for hours at a time. The trackways of prosauropods and sauropods show that, like most dinosaurs, they normally walked at speeds over 3 km/h, much faster than the slow speeds recorded in the trackways of prehistoric reptiles. The dinosaurs' legs and the trackways they made both indicate that the animals' sustained aerobic capacity exceeded the reptilian maximum.

Even the fastest reptiles have slender leg muscles because their low-capacity respirocirculatory systems cannot supply enough oxygen to a larger set of locomotory muscles. Mammals and birds tend to have large leg muscles that propel them at a fast pace over long distances. As a result, mammals and birds have a large pelvis that supports a broad set of thigh muscles. It is interesting that protodinosaurs, the first theropods, and the prosauropods had a short pelvis that could have anchored only a narrow thigh. Yet their legs are long and erect. Such a combination does not exist in any modern animal. This suggests that the small-hipped dinosaurs had an extinct metabolic system, probably intermediate between those of reptiles and mammals. All other dinosaurs, sauropods among them, had the large hips able to support the large thigh muscles typical of more aerobically capable animals.

That many dinosaurs, including all long-necked prosauropods and sauropods, could hold their brains far above the level of their hearts indicates that they had the high levels of power production seen in similarly tall birds and mammals. This would have been especially true of the ultratall sauropod giants. Just how much so is, however, a matter of question, because it is not certain whether extra tall creatures are stuck with just using ever-higher blood pressures to push blood many meters above heart level, or whether they can utilize a siphon effect to partly reduce the vertical loads in the blood column. Some work on giraffes suggests the latter is operative to some level, but further research awaits.

It has long been questioned how high-metabolism sauropods could have fed themselves with the small heads that made their long necks possible. But there is no relationship between head size and metabolism. Herbivorous lizards, mammals, and some birds have large heads relative to their bodies; the largest birds have small heads. The small head of a sauropod was like that of a tachyenergetic emu or ostrich—it was basically all mouth. Most of the large heads of herbivorous mammals consists of the dental batteries they

Giraffatitan brancai

use to chew food after it has been cropped with the mouth, which is restricted to the front end of the jaw. Also, sauropod heads are not as small as they look—that is an illusion created by the size differential with their titanic bodies—with the mouths of the biggest sauropods able to engulf the entire head of a giraffe. Prosauropod mouths were as broad as those of herbivorous mammals or birds at a given body mass, and the breadth of the sauropod mouth is an isometric extension of that line. If a tachyenergetic sauropod of 50 tonnes ate as much as a mammal of its size is expected to eat, then it would have needed to consume over half a tonne of fresh fodder a day. But that is only 1 percent of its own body mass, and if the sauropod fed for 14 hours each day and took one bite per minute, then it would have needed to bite off just half a kilogram (1 lb) or so of plant material each time. That would have been easy for the sauropod's head, which weighed as much as a human body and had a mouth about 0.5 m wide (1.5 ft). Making it all the easier was that the sauropod did not need to take time to give each bit a good chew before swallowing, which it could do almost as fast as it cropped the foliage.

Turning from eating to breathing, an intermediate metabolism is compatible with the unsophisticated lungs that prosauropods appear to have had. The highly efficient, birdlike, air sac–ventilated respiratory complex of sauropods is widely seen as evidence that elevated levels of oxygen consumption evolved in these dinosaurs; no animals with reptilian energetics have such sophisticated respiratory complexes. Sauropods probably needed a bird-like breathing complex in order to oxygenate a high metabolic rate through their long tracheae.

Many birds and mammals have large nasal passages that contain respiratory turbinals. These are used to process exhaled air in a manner that helps retain heat and water that would otherwise be lost during the high levels of respiration associated with high metabolic rates. Because reptiles breathe more slowly, they do not need or have respiratory turbinals. Some researchers point to the lack of preserved turbinals in dinosaur nasal passages, and the small dimensions of some of the passages, as evidence that dinosaurs had the low respiration rates of bradyenergetic reptiles. However, some birds and mammals lack well-developed respiratory turbinals, and in a number of birds the turbinals are completely cartilaginous and leave no bony traces. Some birds do not even breathe primarily through their nasal passages: California condors, for example, have tiny nostrils. The biggest birds have the relatively smallest heads and correspondingly small respiratory passages, in parallel with sauropodomorphs. The turbinal evidence is not definitive.

The low exercise capacity of land reptiles appears to prevent them from being active enough to gather enough food to grow rapidly. In a biological expression of the principle that it takes money to make money, tachyenergetic animals are able to eat the large amounts of food needed to produce the power needed to gather the additional large amounts of food needed to grow rapidly. Tachyenergetic juveniles,

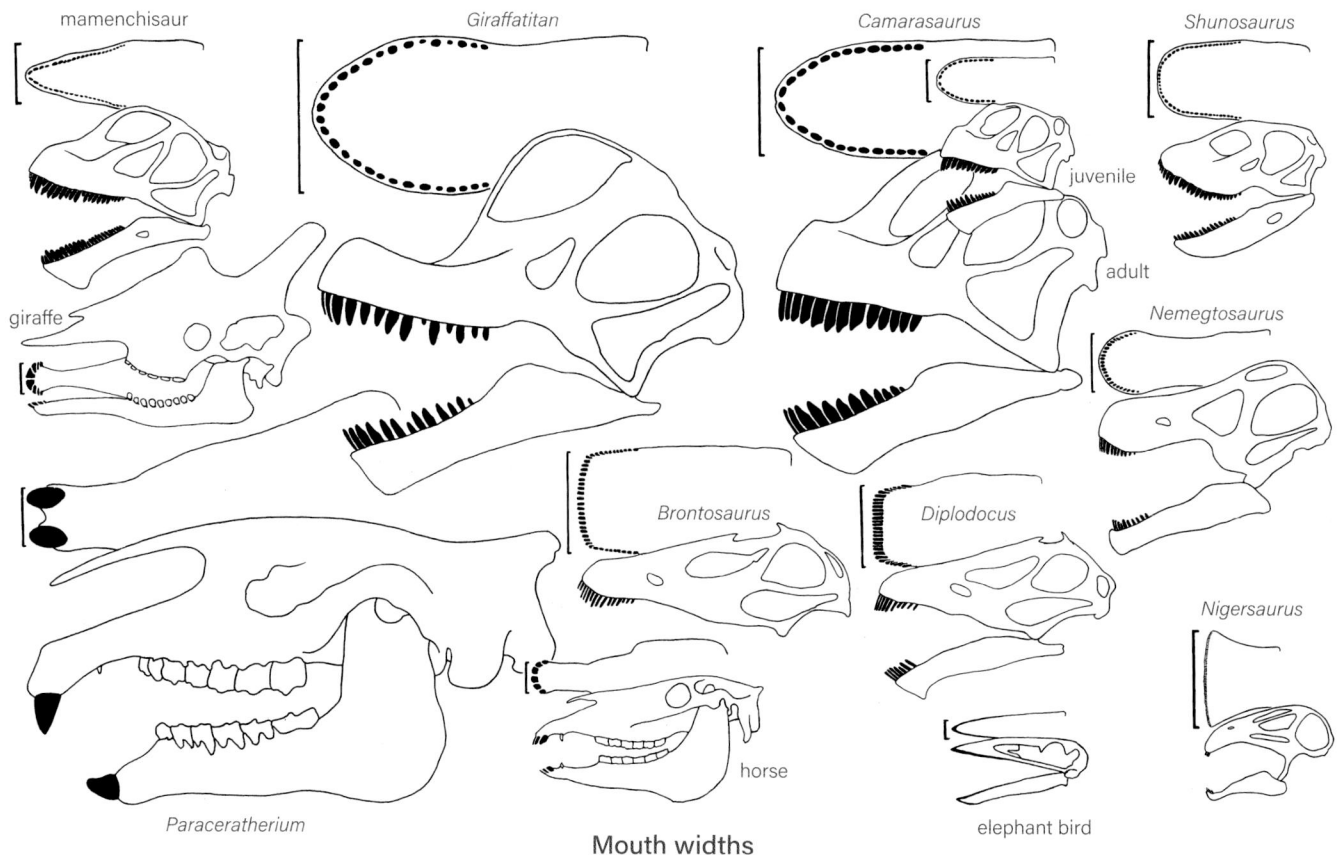

Mouth widths

such as those of sauropods, either gather the food themselves or are fed by their parents in other cases. That sauropodomorphs grew at rates faster than those seen in land reptiles indicates that the former had higher aerobic capacity and energy budgets. It has been suggested that sauropods experienced a radical drop from mammalian to more reptilian energetics as they matured. This would have required an extreme transformation of their cellular biology and organ operations that never occurs in living animals, such probably not being possible. If anything, the increasingly high power output demanded by the increasingly high-pressure respirocirculatory complex needed to get the blood up to the brains ever more stories higher than the heart may have required rising metabolisms as the animal skyscrapers completed growth.

A hot topic has been the long-standing concern by many that the big sauropods would have overheated in the Mesozoic greenhouse if they had had avian or mammalian levels of energy production. However, the largest animals dwelling in the modern tropics, including deserts, are big birds and mammals. And consider that there are no reptiles over a tonne dwelling in the balmy tropics. Further consider that some of the largest elephants live in the Namib Desert of the Skeleton Coast of southwestern Africa, where they often have to tolerate extreme heat and sun without the benefit of shade—they have been observed striding across the sunny, shadeless landscape when the temperature was 38°C (100°F). It is widely thought that elephants use their big ears to keep themselves cool when it is really hot, something dinosaurs could not do. However, elephants flap their ears only when the ambient temperature is below that of their bodies. When the air is as warm as the body, heat can no longer flow out, and flapping the ears actually picks up heat when the air is warmer than the body, so the ear activity tamps down. Nor was the big-eared African elephant the main savanna elephant until fairly recently; the dominant open-area proboscidean used to be one of the biggest land mammals ever, *Palaeoloxodon recki*. A relative of the Asian elephant, it probably had small ears of little use for shedding body heat at any temperature. It is actually small animals that are most in danger of suffering heat exhaustion and heat stroke because their small bodies pick up heat from the environment very quickly. The danger is especially acute in a drought, when water is too scarce to be used for evaporative cooling. Because large animals have a low surface area/mass ratio, they are protected by their bulk against the high heat loads that occur on very hot days, and they can store the heat they generate internally. Put a small dog without water in an open field from which it cannot escape on a hot, sunny day and it will die. An elephant in the same circumstances will not be happy, but it will live. Large birds and mammals retain the heat they produce during the day by allowing their body temperature to climb a few degrees above normal and then dumping it into the cool night sky, preparing for the cycle of the next day. Sauropods may have taken an activity break in the shade on hot middays when possible, but it was not critical. To be avoided was engaging in combat at such times, but the giant theropods would also have stressed themselves out heatwise and were not likely to attack at noon.

At the other end of the temperature spectrum, the presence of a diverse array of dinosaurs in temperate polar regions and highlands that are known to have experienced freezing conditions during the winter, and were not particularly warm even in the summer, provides additional evidence that dinosaurs were better able to generate internal heat than reptiles, which were scarce or totally absent in the same habitats. It was not practical for land-walking dinosaurs to migrate far enough toward the equator to escape the cold; it would have cost too much in time and energy, and in some locations oceans barred movement toward warmer climes. The presence of sauropods in some of the wintry habitats in northern Australia and central Asia directly refutes the hypothesis that big dinosaurs used their bulk to keep warm by retaining the small amount of internal heat produced by a reptilian metabolism; only a higher level of energy generation could have kept the body core balmy and the skin from freezing. That sauropods are missing from the most extreme polar regions is probably because the cold, dark winters left them without enough food to power and warm up their titanic bodies. The sauropods' long necks would have been a particular source of heat loss to the environment not present in other large dinosaurs.

The small, altricial juveniles of prosauropods, stuck in their open nests and exposed to the elements, including cooling rains, would have benefited from, if not needed to have, elevated metabolisms.

The isotopes of chemical elements in bones have been used to help assess the metabolism of dinosaurs. These can be used to examine the temperature fluctuations that a bone experienced during life. If the bones show evidence of strong temperature differences, then the animal was heterothermic on either a daily or seasonal basis. In this case, the animal could have been either a bradyenergetic ectotherm or a tachyenergetic endotherm that hibernated in the winter. The results indicate that dinosaurs large and small were more homeothermic, and therefore more tachyenergetic and endothermic, than crocodilians from the same formations. One study found evidence for increasing metabolic rates with growth in sauropods, which if correct is in accord with the power requirements of their increasing height.

Bone biomolecules, too, are being used to restore the metabolic rates of dinosaurs. This effort is in its early stages, and it is not clear that the sample of living and fossil animals of known metabolic levels is yet sufficient to establish the reliability of the method. And the sample of dinosaurs is also too limited to allow high confidence in the results to date. This is all the more true because the estimates for dinosaurs appear inconsistent in peculiar ways. While the one armored ankylosaur is attributed with a high energy budget that appears excessive for such a relatively slow-moving creature with weak dentition, the sole armored stegosaur is recovered well down in the reptilian range, which looks both too low for an animal with long, erect legs and fairly fast growth and too different from the other armored dinosaur. Also problematic is that reptilian energetics are assigned to the hadrosaur and the ceratopsid examined, not the higher levels expected in animals with such fast food processing and growth that have the large leg muscles and fast walking pace expected in tachyenergetic endotherms. The initial biomolecule results indicating that the earliest dinosaurs were endotherms, with theropods and sauropods great and small remaining so, while some ornithischians irregularly reverted to bradyenergetic ectothermy, await further analysis.

Because the most basal and largest of living birds, the ratites, have energy budgets similar to those of marsupials, it is probable that most dinosaurs did not exceed this limit. This fits with some bone-isotope data that seem to indicate that dinosaurs had moderately high levels of food consumption, somewhat lower than seen in most placentals of the same size. Possible exceptions are the tall sauropods with their high circulatory pressures. It is likely that sauropodomorphs, like birds and some mammals, were less prone to controlling their body temperatures as precisely as do many mammals. This is in accord with their tendency to lay down bone rings. Big sauropods' daily body temperatures should have fluctuated strongly as they stored heat during hot days and off-loaded it into the night sky. Because dinosaurs lived on a largely hot planet, it is probable that most had high body temperatures of 38°C (100°F) or more to be able to resist overheating. The possible exception was again high-latitude dinosaurs, which may have adopted slightly lower operating temperatures and saved some energy, especially if they were active during the winter. Some researchers have characterized dinosaurs as mesotherms intermediate between reptiles, on the one hand, and mammals and birds on the other. But because some mammals and birds themselves are metabolic intermediates, and dinosaurs were probably diverse in their energetics, with some in the avian-mammalian zone, it is not appropriate to tag dinosaurs with a uniform, intermediary label.

A horsepower is the work that can be aerobically sustained without undue fatigue by a large workhorse over a work period of some hours, such as turning a wheel pump or pulling a plow. When going all out anaerobically for a brief period—while pulling a heavy sled in a competition at a country fair, for example, or when a thoroughbred wins the Triple Crown—a horse can do about 15 hp. Male human athletes can put out a third of a horsepower indefinitely, and about 2.5 hp briefly, over twice that of a nonathlete. With energy output scaling to the three-quarters power of mass, a 70 tonne sauropod was capable of producing a maximum of about 500 hp anaerobically and could sustain 30 hp aerobically, similar to that of giant whales.

Until the 1960s, it was widely assumed that high metabolic rates and/or endothermy were an atypical specialization among animals, being limited to mammals and birds,

and perhaps some therapsid ancestors of mammals, and the flying pterosaurs. The hypothesis was that being tachyenergetic and endothermic is too energy expensive and inefficient for most creatures and evolved only in special circumstances, such as the presence of live birth and lactation, or powered flight. Energy efficiency should be the preferred status of animals, especially the big ones, as it reduces their need to gather food in the first place. Since then, it has been realized that varying forms of tachyenergy definitely are or probably were present in large flying insects, some tuna and lamnid sharks, some basal Paleozoic reptiles, some marine turtles, and the oceangoing plesiosaurs, ichthyosaurs, and mosasaurs, brooding pythons, basal archosaurs, basal crocodilians, pterosaurs, all dinosaurs including birds, some pelycosaurs, therapsids, and mammals. Energy-expensive elevated metabolic rates and body temperatures appear to be a widespread adaptation that has evolved multiple times in animals of the water, land, and air. This should not be surprising in that being highly energetic allows animals to do things that bradyenergetic ectotherms cannot do, and natural selection via DNA survival acts to exploit available lifestyles that allow reproductive success without a priori caring whether it is done energy efficiently or not. Whatever works, works. So many animals do live on low, energy-efficient budgets, while others follow the scheme of using more energy to acquire yet more energy that can be dedicated to reproducing the species.

A long-term debate asks what specifically it is that leads animals to be tachyenergetic and endothermic. One hypothesis proposes that it is habitat expansion, that animals able to keep their bodies warm when it is cold outside are better or exclusively able to survive in chilly places—near the poles, at high altitudes, in deep waters—or during frosty nights. Another hypothesis proffers that only tachyenergetic animals with high aerobic capacity can achieve high levels of sustained activity regardless of the ambient temperature, whether at sea level in the tropical daylight or during polar winter nights, and that ability is critical to going high energy aerobic-wise. Certainly, the first hypothesis is true, but it is also true that all of the many animal groups that feature high energy budgets and warmer-than-ambient body temperatures also thrive in warm and even hot climes, where they beat out the bradyenergetic creatures in sustained activity levels. So both hypotheses are operative, and which is more so depends on the biocircumstances—including being really big on land.

THE LARGELY TERRESTRIAL LIFE

Illustrations showing giant, long-necked sauropods snorkeling in deep waters were not viable because water pressure at depth would have prevented inhalation. Whales can exhale as their chests are still deep, but to inhale, the body has to be awash and the lungs near the surface. In any case, sauropods could not stand in deep water for the obvious reason that they were low-density animals due to their pneumatic bones and associated air-sac complex. This is the opposite of hippos, which are so bone-dense that they cannot surface swim; instead, they bottom walk. While most land mammals are nearly awash when swimming and have to power swim to avoid drowning, pneumatic ratites have their upper backs above water, as would have sauropods. Buoyant diving birds have to use powerful propulsive organs and/or do aerial dives to get well underwater; sauropods lacked such. Deep fresh waters are rather rare in any case, and large rivers would be dangerous for deep-water walkers due to their powerful currents. Elephants are good swimmers that enjoy a safety factor by breathing through their trunks, which can be held well out of the water. The more buoyant long-necked sauropods, too, would have been safe swimmers. Those sporting very elongated cervical series may have had to stretch them out nearly horizontal to avoid tipping over or pitching forward. Prosauropods would have been fine swimmers; although not highly buoyant, the long necks of euprosauropods would have protected them from drowning. Adding to the absence of swimming specializations, including for snorkeling, was that sauropod nostrils were apparently not placed atop their heads.

Further deterring the life aquatic for sauropods was the scarcity of underwater sustenance. Flowering angiosperms,

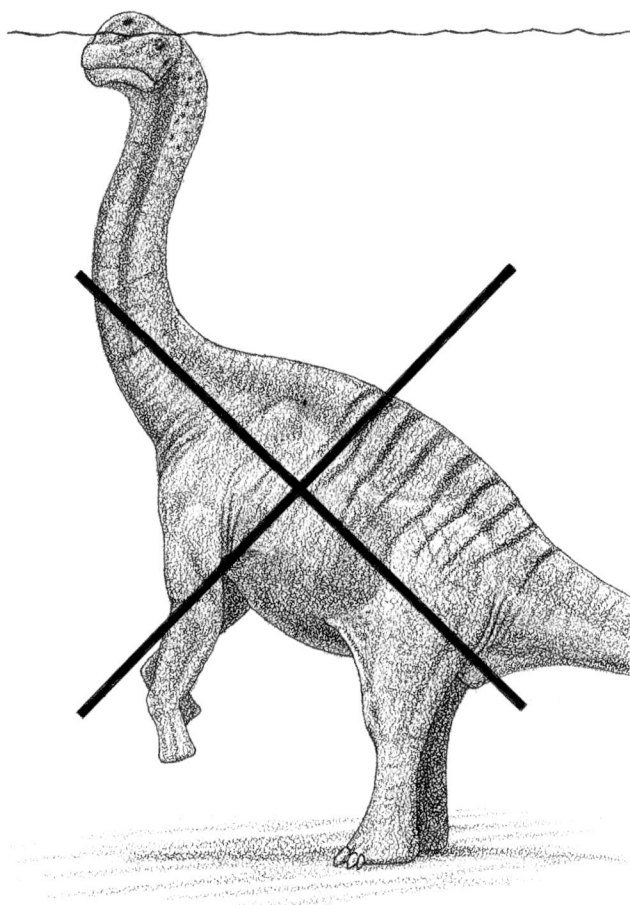

such as water lilies, did not inhabit fresh waters until the Early Cretaceous, and previous water plants were of low caloric and nutritive value. No modern land mammal feeds exclusively on aquatic vegetation; even hippos are land grazers. Deep rivers and lakes lack much in the way of bottom vegetation because light levels are too low. More nutritious plants, such as horsetails and later water grasses, lined Mesozoic waterways, but these could be accessed from firmer ground ashore.

That would have been a good feeding tactic because sauropod feet were poorly suited for aquatic habits. The hindfeet were largely large, rounded, toeless pads like those of elephants, and the latter are predominantly terrestrial. The dinosaurs' hands were worse, being unpadded pillar structures with little or nothing in the way of splayed fingers. In sauropod trackways, the handprints are normally deeper than those of the hindfeet despite the hindlimbs bearing more weight than the fore. That was because the pliable hindpad spread out the load over a large area, while the much smaller hands plunged deeper into the muds and sands. Had sauropods been water lovers, they would have had very different appendages. Some hand-only sauropod trackways may be those of long-armed examples poling along with the arms on the bottom while floating in sufficiently shallow waters. But with the lung-air-sac complex largely in the neck and chest area, sauropods would have been hip- and tail-heavy. At least some of the hand-only tracks are sauropods walking on sediments too firm to capture the large-area hindfeet prints but soft enough for the narrower hands to penetrate—or they are underprints in which only the deeper foreprints remain.

Because sauropods were seen as riverine and lacustrine animals, it was largely presumed that their ancient habitats were well-watered lands featuring major twisting rivers and clear-water lakes, rather like the Amazon basin. The geology of sauropod-bearing formations indicates otherwise. The rivers were often shallow, braided, and seasonally fluctuating, and lakes, if present, were likewise shallow and sometimes alkaline and saline. Sauropods appear to have been especially fond of formations that feature extensive carbonite caliche deposits. These need a combination of substantial wet-season rainfall, followed by soil-drying dry seasons, to form. In sauropod times, these would have been open, forested areas with riverine vegetation well suited for the high-browsing activities of tall sauropods. These lands would have had extensive fern cover between the trees that sauropods able to low browse could have exploited, at least in the wet seasons. During the dry period, these sauropods could have gone tripodal to dine on tree fodder that would have been sustained by the deep roots tapping into groundwater. Other sauropod habitats were more arid, with trees and underbrush concentrated along watercourses; this is similar to the habitats of the desert elephants, such as those of the Skeleton Coast. Other sauropod habitats look like they were rainier than the above.

Coelophysis rhodensis and
Massospondylus carianatus

Substantial populations of large herbivores have a significant impact on the floras they feed on. Consider the browse lines that a modestly sized prosauropod would have left at heights up to 6 m (20 ft) or so. A 50 tonne sauropod would consume in the area of 200 tonnes of forage a year. Over 60 years, that would be about 12,000 tonnes, the weight of a World War II cruiser. The dinosaur would return about a third of that as fertilizing feces. Savanna elephants can extra dramatically alter landscapes by pushing down trees too tall to allow access to their upper foliage yet small enough to topple. Bigger, and with clawed forelimbs on hand, sauropods would have been able to fell larger trees as desired. During droughts seasonal or otherwise, sauropodomorphs could have used their hindclaws to dig for water—which other animals might have tapped into. This may help explain why the sauropods retained toe claws until their end-Mesozoic demise.

DEFENSE

Adult prosauropods and sauropods were, like many herbivores, well armed and able to attempt to defend themselves when attacked. This would have been necessary because they had no hope of outrunning the bird-limbed theropods. Nor was hiding a viable option for giant sauropods that were prone to making considerable noise and presumably stank, like large herbivorous mammals—their predators would likely have been less odorous, allowing them to sneak up on them. On the other hand, the great dimensions of sauropods were a defense positive in that having a set of high-performance eyes with in the area of 360 degrees of vision held stories above ground level would have often aided in spotting danger at long distances, an ability boosted by rearing up when it seemed necessary. No other animals would have had this long-range predator detection advantage. Trying to stay out of sight was more viable for prosauropods, the lesser sauropods, and juvenile sauropods, although they, too, in some cases benefited from being able to stand tall to see approaching danger.

Having some sharp teeth, early prosauropods could have bitten at attackers when unable to dodge attack. Prosauropods could have lashed out with their large thumb and toe claws, using both sets while sometimes bouncing on their tails kangaroo-style. Sauropod battle qualifications usually but not always included sheer size, enormous tails up to 10–15 m long (35–50 ft), and weighing between 3 and 20 tonnes—often the total mass or more of the theropods attacking them—and, except for titanosauriforms, big, stout thumb claws. Why the latter were reduced and then lost in the last of the sauropods is a mystery. Swinging tails that among diplodocoids and titanosaurids were tipped with long, lashing whips that some believe could achieve sonic-crack-producing supersonic speeds, or with small tail clubs in some examples, and rearing to deploy hand claws would have posed stout defenses against even the biggest theropods. Hand-claw defense is used by anteaters, sloths, and probably the extinct giant ground sloths and the chalicothere horse relatives. Being bipedal up on two legs as per retroverted hipped camarasaurs would have had a mobility advantage over the more static tripodal stance of other sauropods. Primary sauropodomorph vulnerabilities would have been the leg retractors of the tail base and thighs and the slender, delicate neck, at least among those small enough for their neck base to be reached. The placement of

Early Jurassic *Coelophysis* and *Massospondylus*

sauropods' cervical arteries and veins up between the neck ribs would have offered the critical blood-flow vessels a fair amount of protection from attack. Big sauropods may have been at risk of avepods dashing in, slicing or punch biting out a chunk of flesh, and then darting away before an effective counterattack was made, and doing so on a repeat basis until satisfied. The avepod's aim would not have been to outright kill the land titan, but the resulting wounds may have been debilitating and ultimately lethal, in which case the vast carcass would have been a huge smorgasbord for the local avepods until it was consumed or rotted. Predators can gorge themselves by about a quarter of their body mass, so they could have chowed down on the carcass for two or three days and gone for a dozen days on that. A deceased 20 tonne sauropod would be ready to eat: over the next few days, three dozen 1.5 tonne *Allosaurus* would have come in to take their portion before the sauropod decayed. Each would have gotten a full meal that fueled them two weeks, given that metabolic rates per unit of body mass decline with large body mass, so larger avepods could have gone longer between meals.

A famous Early Cretaceous Texas trackway records a sauropod approaching two dozen tonnes that is closely paralleled on its immediate left by an allosauroid of a few tonnes. That they were moving in synch at about the same speed, and that when the sauropod tracks bend a little to the left, so do those of the avepod, indicate that the similar paths were not coincidental. When the direction shift occurs, the sauropod appears to have stumbled, while the predator seems to have missed a step. This indicates the latter attacked at the moment, possibly slashing at the

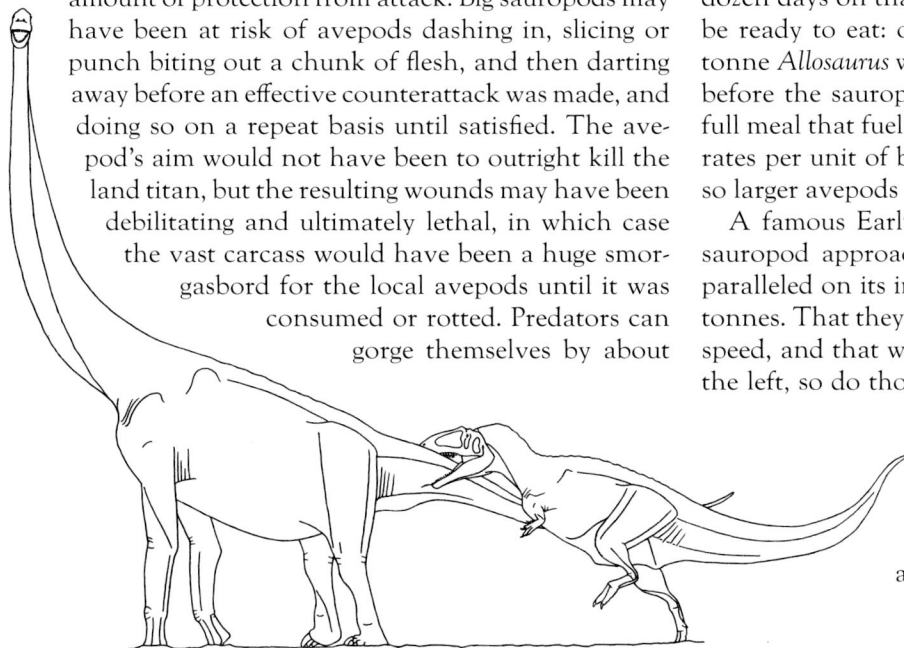

Giant avepod attacking sauropod trackway

Yangchuanosaurus shangyouensis and unnamed genus *hochuanensis*

tail-based leg retractor muscles in order to try to cripple its target as part of a killing process. While latched onto its prey, the avepod was pulled along and misstepped, while the pained victim was pushed off balance. That the herbivore slowed down after the critical points suggests the attack was successful. Skeptics question the statistical likelihood that such an event would happen to be preserved, but the particular attack was probably one among many as the carnivore worked away at damaging its victim enough to kill it, perhaps over a period of hours and a few kilometers' distance.

Another tactic of desperation is to run into a body of water. This brings us to the conceit that was once the conventional wisdom—that dinopredators were hydrophobic to the degree that all an herbivore under pursuit had to do was go for a swim and leave the vexed theropod standing frustrated on the shoreline. The thin premise was that the narrow toes of theropods left them more prone to get mired or rendered them poorer swimmers. When it was thought that sauropods were predominantly aquatic, this was seen as their go-to predator defense. This water-escape notion has largely fallen by the wayside with the realization, based partly on the bottom-poling avepod trackways, that dinopredators were of course adept swimmers capable of pursuing their victims into water. And mammalian carnivores are known to chase down panicked mammalian herbivores that try the river or lake to escape.

But the water trick should not be dismissed out of hand. If a dinosaurian carnivore lived in Late Triassic environs in which watercourses were infested by very large crocodilian-like phytosaurs, or Cretaceous supercrocodilians up to 10 tonnes, then dashing into waters graced by such terrestrial-beast-drowning monsters would have been dangerous—and for the prey target, too, the water option may have been leaping from the frying pan into the fire. The exception being titanic sauropods. Fear of crocs may be why carnivorous mammals do not always chase game into tropical waters. And there may be no point for land predators to kill prey in crocodilian-dominated waters because the crocs will happily take over the carcass conveniently floating in the habitat in which they have the advantage. Only if a big dinosaur can quickly carry or drag its victim ashore and out of reach of crocodilian jaws is it advisable to dispatch it in the latter's territory in the first place. If the prey dinosaur was a nonpneumatic prosauropod that either lived before the age of big crocs or was willing to take the risk posed by Triassic and Cretaceous aquatic archosaur predators, resorting to water could be a good option for outfoxing a theropod. The latter would be too buoyant to dive after a nonpneumatic dinosaur if it dove beneath the surface in sufficiently deep water, where the theropod would not even be able to track it. The pneumatic sauropods could not do the deep-dive trick even when juveniles.

Titanosaurs were the only armored sauropodomorphs. What the osteoderms were really for is not clear. Although the specific arrangements are not preserved, it is clear that the armor did not form a dense protective cover like that of ankylosaurs, armadillos, or the latter's extinct glyptodont relations. Whether titanosaur armor would have been effective is not clear, but it may have improved the superficial protection somewhat. Titanosaurs faced the greatest supertheropod threat of the Sauropoda, initially in the form of giant allosauroids, so some extra fortification would have been helpful. The alternative hypothesis sees the skin bones as calcium reserves to aid rapid production of large numbers of big eggs that only titanosaurs among sauropods are so far proven to have produced, but this does not preclude other functions.

An advantage of nest care by the smaller sauropodomorphs would have been protection of the nestlings that otherwise would have been roaming on their own. Not being watched over, the hatchlings of sauropods would have been easy meals for small predators. In that case, rapid reproduction would make up for the heavy losses.

GIGANTISM

Although dinosaurs evolved from small protodinosaurs, and many were small—birds included—dinosaurs are famous for their tendency to develop gigantic forms. The average mammal is the size of a dog, whereas the average fossil dinosaur was bear-sized. But those are just averages. Predatory theropods reached as much as 10 tonnes or more, as big as elephants and dwarfing the largest carnivorous mammals by a factor or 10 or more. A number of sauropods exceeded the size of the largest land mammals, mammoths, and the long-legged paracerathere rhinos of 15–20 tonnes by a factor of at least four to five and apparently matched the most massive whales.

Among land animals whose energetics are known, only those that are tachyenergetic have been able to become gigantic on land. The biggest fully terrestrial reptiles, some oversized tortoises and monitors, have never much exceeded a tonne. Land reptiles are probably unable to grow rapidly enough to reach great size in reasonable time. Other factors may also limit their size. It could be that living at 1 g, the normal force of gravity, without the support of water, is possible only among animals that can produce high levels of sustained aerobic power. The inability of the low-power, low-pressure reptilian circulatory system to lift blood far above the level of the heart probably helps limit the size of bradyenergetic land animals. That a number of Mesozoic dinosaurs exceeded a tonne, as have mammals since then, is compelling evidence that they, too, had high aerobic power capacity and the correspondingly elevated energy budgets. The ultimate example of great height driven by elevated metabolics is seen in ultratall

blue whale, 140–200 t

Palaeoloxodon, 12–20 t

African elephant, 6–10 t

Paraceratherium, 8–17 t

Brontosaurus (20 t)

Bruhathkayosaurus, 130–200 t

therapsid *Lisowicia*

Argentinosaurus, 75–80 t

unnamed genus *youngi* (35 t)

Plateosaurus (2 t)

Patagotitan, 60 t

Maraapunisaurus, 80–120 t

ground sloth *Eremotherium* 4 m

4 meters

Xinjiangtitan, 40 t

Giraffatitan, 35 t

giraffe, 1.5 t

Dinosaur giants compared with mammals

sauropods. Their extreme height indicates that their hearts could push blood many meters up against the gravity well at pressures up to two or three times higher than the 200 mm Hg giraffes need to oxygenate their brains. And it is unlikely that such tall and massive animals in danger of fatal injury from falling could risk a moment of hypoxic wooziness from an oxygen-deprived brain. If so, then sauropods had extra hardworking hearts whose high energy demands would have required a very high level of oxygen consumption.

Only sauropods have exceeded 20 tonnes on land. The question is, why that unique biological achievement? Very tall necks like those of sauropods and giraffes evolve in an evolutionary feedback loop that involves two distinct but reinforcing factors. Increasing height serves as a dominance display that enhances reproductive success by intimidating rivals and impressing mates. This is similar to other reproductive displays, such as the tails of peacocks and the giant antlers of big cervids. And as the head gets higher, the herbivore also has a competitive feeding advantage over shorter herbivores in accessing the enormous food resources, in the crowns of tall trees, that provide the power source needed to pump blood to the brain, held far above the heart, that allows the animal to reach all that food. Sauropods could take this to an exceptional extreme because, lacking dental batteries and big brains, sauropod heads were relatively small—and because of their extensive sinuses, low density—so sauropods were able to evolve extremely tall necks that in turn required enormous bodies to anchor them upon and to contain the hardworking hearts they needed. With their toothier big heads, mammals are apparently limited to the 6 m (20 ft) height of giraffes. The tallest sauropods were able to reach up to maybe 20 m (60 ft)—without better understanding of how animals get blood really high up, it is not known whether that was the maximum animal height attainable. Not having such pressure problems, and able to use capillary action to draw water upward, trees can exceed 100 m (350 ft). Toward the other end of the size spectrum, those sauropods specialized with broad, squared-off mouths for grazing ground cover did not need supernecks or massive bodies to support them, and were relatively lightweight rhino-sized creatures. This disproves the belief that sauropod necks were long to improve low browsing—necks many meters long are so expensive to grow and maintain that they evolve only under strong, practical selective pressures that provide advantages short necks could not. It was otherwise-unreachable floral heights that made most sauropods tall and titanic. Very long necks can actually hinder grazing. They have to sweep back and forth over a wide area. That may sound like a good idea, but tall bushes and trunks scattered about the landscape, whether it be hills, flats, or shorelines, may frequently get in the way of the appendage; similarly, short-gunned tanks can have an advantage in wooded areas, where those with long weapons get caught up in the foliage. Herbivores specialized for feeding on low-lying flora don't need or benefit from long, awkward, and costly necks; they simply walk up to the plant, reach down, and eat it.

Also possibly helping those sauropods that did become supersized were the pneumatic vertebrate and other air sacs that had evolved to improve respiratory capacity in tune with the high metabolic rates needed to be so tall and heavy. Those lightened the load on their bones and muscles somewhat, which could have been especially pertinent regarding the elongated necks. This option has not been available to mammals, or to ornithischian dinosaurs, for that matter. But this effect should not be exaggerated, in part because recent work is showing that sauropods' internal air spaces were not as dramatically density reducing as has been widely thought.

As is typical of terrestrial vertebrates, sauropod skeletons made up 15 percent of their total mass or a little more. The femurs of the larger examples would have weighed hundreds of kilograms; those of humans are in the area of a third of a kilogram. Like proboscideans and whales, sauropods of 10–100 tonnes or more would have consisted of over a quadrillion to tens of quadrillions of cells, compared to 50 trillion or so in a human.

The hypothesis that only tachyenergetic animals can grow to enormous dimensions on land is called terramegathermy. An alternative concept, gigantothermy, proposes that the metabolic systems of giant reptiles converge with those of giant mammals, resulting in energy efficiency in all giant animals. In this view, giants rely on their great mass, not high levels of heat production, to achieve thermal stability. If gigantothermy were true, then reptiles would be as big as mammals on land, but this is not so. This idea of gigantothermy reflects a misunderstanding of how animal power systems work. A consistently high body temperature does not provide the motive power needed to sustain high levels of activity; it merely allows a tachyenergetic animal, and only an animal with a high aerobic capacity, to sustain high levels of activity around the clock. A gigantic reptile with a high body temperature would still not be able to remain significantly athletic for extended periods. And the metabolic rates and aerobic capacity of elephants and whales are as high as expected in mammals of their size and are far higher than those of the biggest crocodilians and turtles—the gigantothermy hypothesis was originally founded on some errantly high measurements of resting metabolic rates of big leatherback sea turtles—which, in fact, have the low levels of energy production typical of reptiles. Also pushing animals to be big is improved thermoregulation. The ratio of high bulk to relatively low surface area makes it easier both to retain internal warmth when it is chilly and to keep external heat out and store internal heat on hot days.

Another, subtle reason that dinosaurs, particularly supersauropods as well as supertheropods, could become so enormous has to do with their mode of reproduction. Because big mammals are slow-breeding K-strategists that

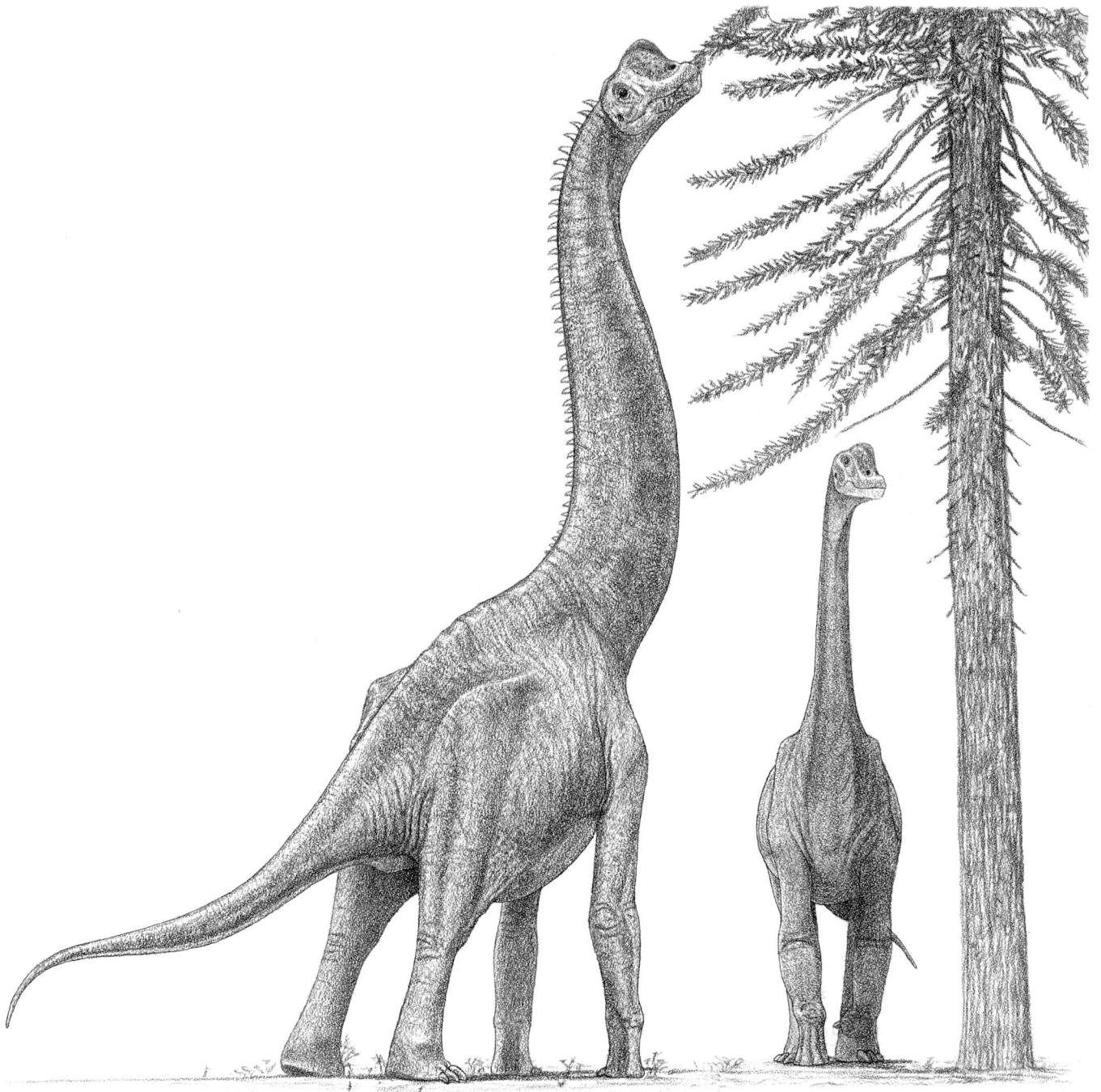

Island dwarfed *Europasaurus*

lavish attention and care on the small number of calves they produce, there always has to be a large population of adults present to raise the next generation. A healthy herd of elephants has about as many breeding adults as it does juveniles, which cannot survive without parental care. Because there always has to be a lot of grown-ups, the size of the adults has to be limited in order to avoid overexploiting their habitat's food resources, which will cause the population to collapse. This constraint appears to limit slow-reproducing mammalian herbivores from exceeding 10–20 tonnes. Because sauropods were fast-breeding r-strategists

that produced large numbers of offspring that could care for themselves, their situation was very different from that of big mammals. A small population of adults was able to produce large numbers of young each year. Even if all adults were killed off on occasion, their eggs and offspring could survive and thrive, keeping the species going over time. Because dinosaurs could get along with smaller populations of adults, the grown-ups were able to grow to enormous dimensions without overexploiting their resource base. This evolutionary scheme allowed plant-eating dinosaurs to grow to over 20 tonnes, perhaps on occasion 200 tonnes. It is notable that supersauropod fossils are particularly rare, indicating small populations. Because the bulk of the biomass of adult herbivorous dinosaurs was tied up in oversized giants, the theropods needed to evolve great size themselves in order to be able to fully access the nutrition tied up in the huge adults—the idea that theropods grew to 6–10 tonnes only to "play it safe" by consistently hunting smaller juveniles is not logical—and the fast-breeding and fast-growing predators could reach tremendous size. The existence of oversized predators in turn may have resulted in a size race, in which sauropods evolved great size in part as protection against their enemies, which later encouraged the appearance of supersized theropods that could bring them down.

It is as common as it is seemingly logical to think that being titanic in water—it being buoyant—is easier than dwelling under the full force of planetary gravity on land, but paleohistory indicates otherwise. Really big whales did not exist until very recently. For over 30 million years, most or all whales were not all that big, rarely if ever exceeding about 40 tonnes. It was under the unusual conditions of ice age oceans, in which cold water currents spurred higher levels of food productivity, plankton particularly, that the superbaleen whales of 50–200 tonnes suddenly appeared. Back in the warm seas of the Mesozoic, reptiles and fish reached only 20 to perhaps 30 tonnes. Before the Pleistocene ocean chill, the biggest fish were 50 tonne supersharks like megalodon. In comparison, 50–200 tonne supersauropods were wandering about Mesozoic continents from the Late Jurassic to the very end of the era with little or no interruption for about 90 million years. The extra tall land giants had the size-sustaining advantage that there were always forests full of calorie-packed big trees, a few trillion of them, on which to grow titanic—until K/Pg events

suddenly wrecked those forests and the equitable climate across the globe. With no small-headed dinosaurian-style herbivores to regain whalelike bulk and multistory height on land, land mammals have proven unable to reach a couple dozen tonnes. How big land animals can get in structural terms is not known. That the femurs of some the largest fat-bellied titanosaurs are on the slender side implies the monsters were not necessarily pushing the terrestrial mass limits.

In the 1800s, Edward Cope proposed what later became Cope's Rule: the tendency of animal groups to evolve gigantism in order to exploit the advantages that accrue from being supersized. The propensity of small-headed sauropods to take this evolutionary pattern to a logical evolutionary extreme means that the Mesozoic saw events on land that are today limited to the oceans. In modern times, combat between giants occurs between orcas and whales. In the dinosaur era, it occurred between orca-sized theropods and whale-sized sauropods.

While sauropods are famed for being very big, those most specialized for grazing ground cover were not, and examples dwelling on islands with limited food resources were prone to island dwarfing—this was similar to some marine-isolated mammoths that were as little as a small pony—but no known sauropods were that miniaturized, although some may have been.

The adult size variation among sauropodomorphs ranging from less than 5 kg (10 lb) to 100 tonnes and up is 20,000-fold or more, an impressive value but well under the 20 million-plus differential seen in mammals, the smallest of which are less than a mere 2 g (.06 oz). Among prosauropods, the variability was a few hundred-fold; for sauropods, it was in the area of 100, although the smallest known examples were the island dwarfs.

MESOZOIC OXYGEN

Oxygen was absent from the atmosphere for much of the history of the planet, until the photosynthesis of single-celled organisms and plants built up enough oxygen to overwhelm the processes that tend to bind it to various elements, such as iron. Until recently, it was assumed that oxygen levels then became stable, making up about a fifth of the air for the last few hundred million years. Of late, it has been calculated that oxygen has fluctuated strongly since the late Paleozoic. The problem is that the results are themselves variable. They do agree that the oxygen portion of the atmosphere soared to about a never-seen-again third or more during the late Paleozoic, when the great coal forests were forming and, because of the high oxygen levels, often burning. It is notable that this is when many insects achieved enormous dimensions by the standards of the group, including dragonfly relatives with wings over 0.5 m (2 ft) across. Because insects bring oxygen into their bodies by a dispersed set of tracheae, the size of their bodies may be tied to the level of oxygen.

But in the Mesozoic, the situation is less clear. Oxygen levels may have plunged precipitously, sinking to a little over half the current level by the Triassic and Jurassic. In this case, oxygen availability at sea level would have been as poor as it is at high altitudes today. Making matters worse were the high levels of carbon dioxide. Although not high enough to be directly lethal, the combination of low oxygen and high carbon dioxide would have posed a serious respiratory challenge that could have propelled evolution of the efficient air-sac respiration of pterosaurs and some dinosaurs. But other work indicates that oxygen did not plunge so sharply, and in one analysis, it never even fell below modern levels, being somewhat higher in most of the era. Reliably assaying the actual oxygen content of the atmosphere in the dinosaur days remains an important challenge.

SAUROPODOMORPH SAFARI

Assume that a practical means of time travel has been invented, and *The Princeton Field Guide to the Prosauropod and Sauropod Dinosaurs* in hand, you are ready to take a trip to the Mesozoic to see the sauropodomorphs' world. What would such an expedition be like? Here we ignore the classic time paradox issue that plagues the very concept of time travel: What would happen if a time traveler to the dinosaur era did something that changed the course of events to such a degree that humans never evolved to travel back in time and disrupt the time line in the first place?

One difficulty that might arise could be the lack of modern levels of oxygen and/or extreme greenhouse levels of carbon dioxide (which can be toxic for unacclimated animals), especially if the expedition traveled to the Triassic or Jurassic. Acclimation could be necessary, and even then, supplemental oxygen might be needed at least on an occasional basis. Work at high altitudes would be especially difficult. But, as noted earlier, oxygen deprivation may not have been the case. Another problem would be the high levels of heat chronically present in most dinosaur habitats. Relief would be found at high latitudes, as well as on mountains.

If the safari went to one of the classic Mesozoic habitats that included gigantic dinosaurs, the biggest problem would be the sheer safety of the expedition members. The bureaucratic protocols developed for a Mesozoic expedition would emphasize safety, with the intent of keeping the chances of losing any participants to a bare minimum. Modern safaris in Africa require the presence of a guard armed with a rifle when visitors are not in vehicles in case of an attack by big cats, hippos, buffalo, rhinos, or elephants. Similar weaponry is often needed in tiger country, in areas with large populations of grizzlies, or in Arctic areas inhabited by polar bears. The potential danger level would be even higher in the presence of flesh-eating prosauropod and sauropod killers as big as rhinos and elephants and able to run down a human who could potentially be out of breath because of the low oxygen. It is possible that theropods would not recognize humans as prey, but it is at least as likely that they would, and the latter would have to be assumed. Aside from the desire not to kill members of the indigenous fauna, rifles, even automatic rapid-fire weapons, might not reliably bring down a 5 tonne allosauroid or tyrannosaur, and heavier weapons would be impractical to carry about. Nor would the danger come from just the predators. A herd of whale-sized sauropods would pose a serious danger of trampling or impact from tails, especially if they were spooked by humans and either attacked them as a possible threat or stampeded in their direction. Sauropods would likely be more dangerous than elephants, whose high level of intelligence better allows them to handle situations involving humans.

But there would be another danger that would be as small as it is big: microbes. Expedition members would be at risk of picking up exotic Mesozoic disease organisms to which they would not be immune, and at least as bad would be the danger of contaminating the ancient environment with a host of late Cenozoic viruses, bacteria, and parasites that could seriously disrupt Jurassic and Cretaceous life.

The combined menaces, small and big, would mean that time-traveling dinosaur watchers would probably be banned from directly interacting with the ancient habitats. Instead of walking about under the Mesozoic sun and stars, breathing ancient fresh air, binoculars in hand, they would always have to wear microbe-proof biohazard suits when not in vehicles, and habitats would likewise need to be sealed against microbes getting in or out. Dwelling in dinosaur habitats would be a lot like living on the moon or Mars—a very artificial experience in which paleonauts would be significantly detached from the fascinating world around them, always respiring pretreated air. An advantage of being in biosuits would be temperature control, which would eliminate dealing with the extreme heat prevalent in much of the Mesozoic. Also dealt with would be issues with the composition of the atmosphere. Travel by foot would probably be largely precluded in habitats that included big theropods, sauropods, and ceratopsids. Expedition members would have to move about on the ground in vehicles sufficiently large and strong to be immune from attacks by colossal dinosaurs. Movement away from the vehicles would be possible only when drones showed that the area was safe. Even in places lacking giant dinosaurs, there would be the peril of a biosuit being breached by an assault by a smaller dinosaur—any such penetration from any cause would require some level of medical care, quarantine at least. Defensive weapons might be necessary, although pepper-spray guns might suffice. Yet another danger in some Cretaceous habitats would be elephant-sized crocodilians that might snap up and gulp down whole a still-living human unwary enough to go near or in the waters where dinosaurs hung out.

A safe way to observe the prehistoric creatures would be remotely via drones. Manned ultralights would work, too, although they would have their own dangers.

A consequence of time travel for paleoartists would be the obsolescence of every single prosauropod and sauropod dinosaur life restoration.

IF TITANOSAURS HAD SURVIVED

Assume that the K/Pg impact is what killed off the titanosaurs, but also assume that the impact did not occur and that titanosaurs continued into the Cenozoic. What would the evolution of land animals have been like in that case?

Although much will always be speculative, the titanosaurs would have remained a force to reckon with—indeed, the Mesozoic era would have endured—aborting the Cenozoic Age of Mammals. Thirty million years ago, western North America probably would have been populated by great dinosaurs, rather than the rhino-like brontotheres. Having plateaued out in size for the last half of the Mesozoic, titanosaurs would probably not have gotten bigger, but the continuation of the ultimate browsers should have inhibited the growth of dense forests. Even so, the flowering angiosperms would have continued to evolve and to produce a new array of food sources, including well-developed fruits that titanosaurs would have needed to adapt to in order to exploit. Even so, it is not knowable if titanosaurs would have made it into more recent times—evolution does not play favorites.

What is also not certain is whether mammals would have remained diminutive or would have begun to compete with dinosaurs for the large-body ecological niches. By the end of the Cretaceous, sophisticated marsupial and placental mammals were appearing, and they may have been able to begin to mount a serious contest with dinosaurs as time progressed. Eventually, southward-migrating Antarctica would have arrived at the South Pole and formed the enormous ice sheets that act as a giant air-conditioning unit for the planet. At the same time, the collision of India and Asia, which closed off the once-great Tethys Ocean and built up the miles-high Tibetan Plateau, also contributed to the great planetary cool-off of the last 20 million years that eventually led to the current ice age despite the long-term rising heat production of the sun. This should have forced the evolution of specialized titanosaur grazers with the broad, squared-off mouths, and teeth that turned over fast to cope with the abrasive grasses of the savannas, steppes, and prairies that thrive in modern climates. Because titanosaurs had already generated ground-cover consumers, doing such should not have been a major difficulty. Not needing to reach 5–20 m (16–66 ft) up into the crowns of trees, the grazing sauropods would likely have been the size of a rhino or small elephant. In terms of thermoregulation, dinosaurs should have been able to adapt to chillier climes, although growing winter food shortages may have been a problem for supersauropods. And the also energetic mammals may have been able to exploit the decreasing temperatures. Perhaps big mammals of strange varieties would have formed a mixed dinosaur-mammal fauna, with the former perhaps including some big birds.

TITANOSAUR CONSERVATION

If we take the above scenario to its extreme, assume that some group of smart dinosaurs or mammals managed to survive and thrive in a world of dinosaurs and became intelligent enough to develop agriculture and civilization as well as an arsenal of lethal weapons. What would have happened to the titanosaur fauna?

The fate of titanosaurs would probably have been grim. We actual humans were the leading factor in the extinction of a large portion of the megafauna that roamed much of Earth toward the end of the last glacial period, and matters continue to be bad for most wildlife on land and even in the oceans. If whale-sized herbivorous dinosaurs were still extant, their low populations likely would have made them more vulnerable than elephants and rhinos. Their eggs would have made for fine dining. By the time the sapients developed industry, the gigantic plant eaters would probably already have been part of historical lore. If superdinosaurs had instead managed to survive in an industrial world, they would have posed insurmountable problems for zoos. Housing and feeding rhinos, hippos, and elephants is not beyond the means of many zoos, but how could a zoo staff handle one or several sauropods that were 12 m (40 ft) tall or taller, weighed 5–50 tonnes or more, and ate up to 10 times as much as an elephant?

WHERE SAUROPODOMORPHS ARE FOUND

Because prosauropods and sauropods are long gone and time travel probably violates the nature of the universe, we have to be satisfied with finding the remains they left behind. With the possible exception of very high altitudes, sauropodomorphs lived in all places on all continents, so where they are found is determined by the existence of conditions suitable for preserving their bones and other traces, eggs and footprints especially, as well as by conditions suitable for finding and excavating the fossils. For example, if a sauropodomorph habitat lacked the conditions that preserved fossils, then that fauna has been totally lost. Or, if the fossils of a given fauna of sauropodomorphs are currently buried so deep that they are beyond reach, then they are not available for scrutiny.

Late Triassic (Rhaetian–Norian–Carnian)

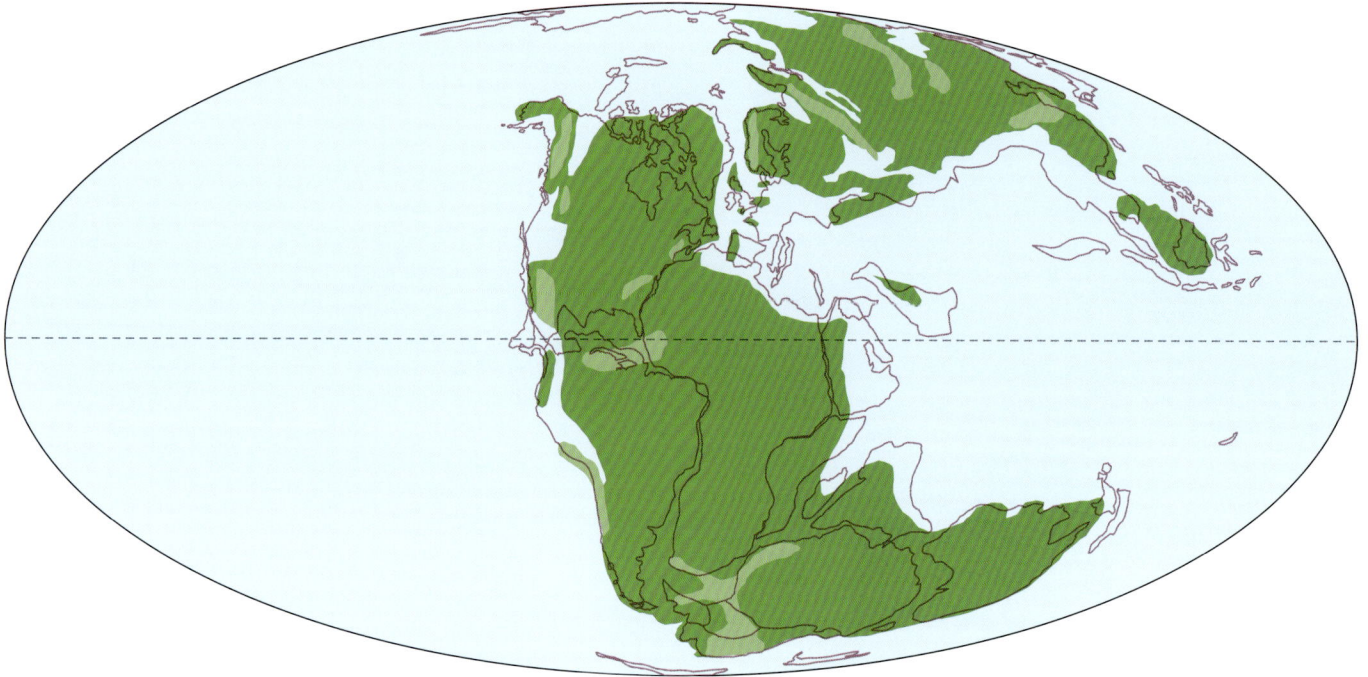

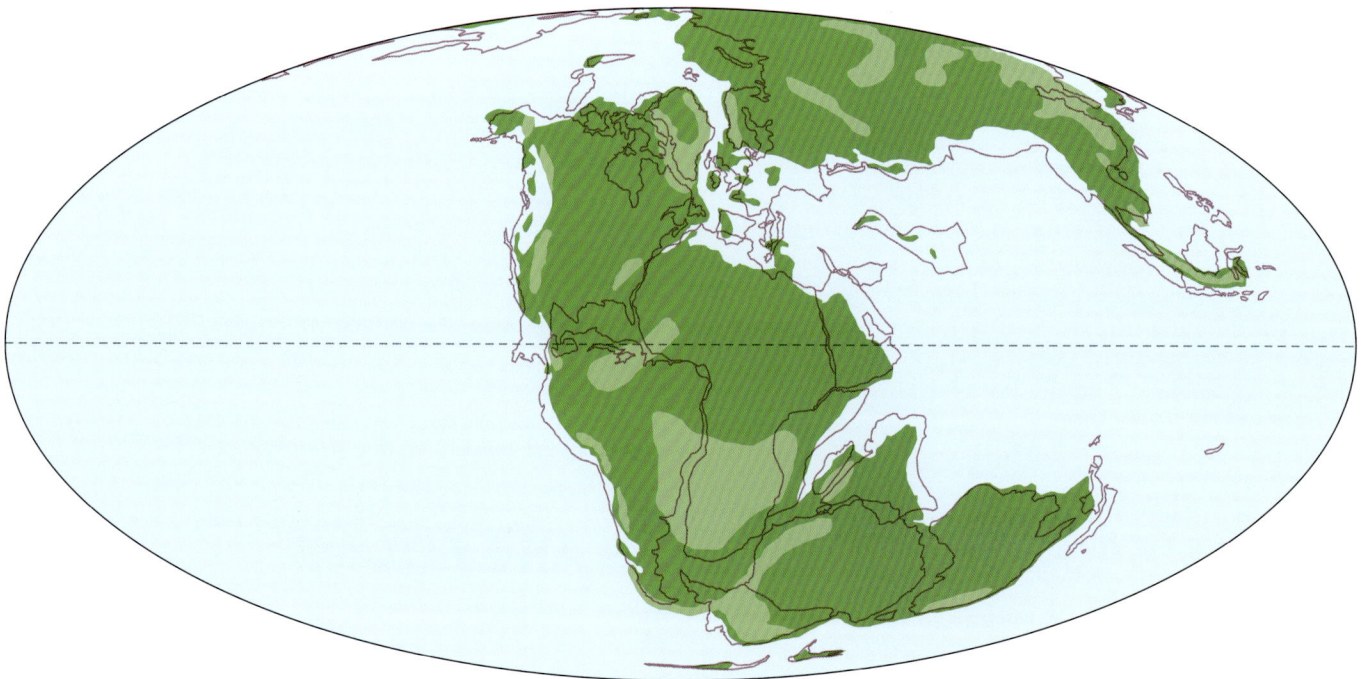

Early Jurassic (Sinemurian)

Middle Jurassic (Callovian)

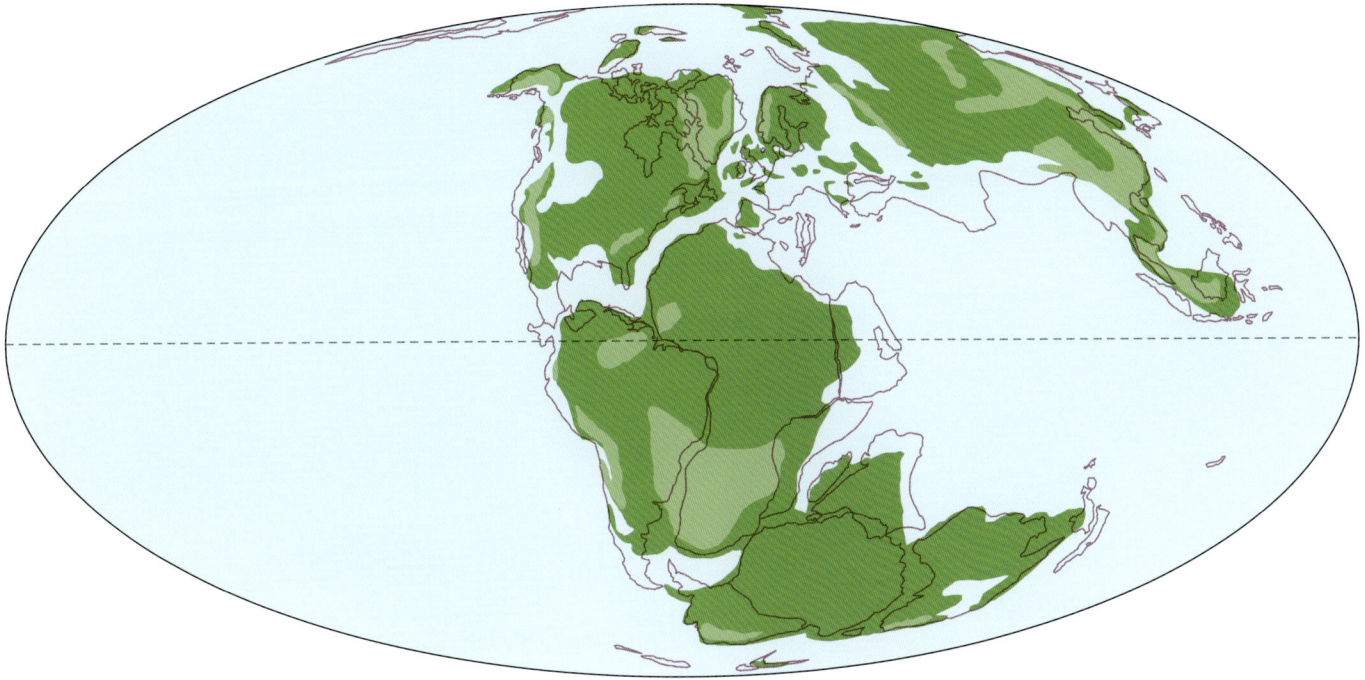

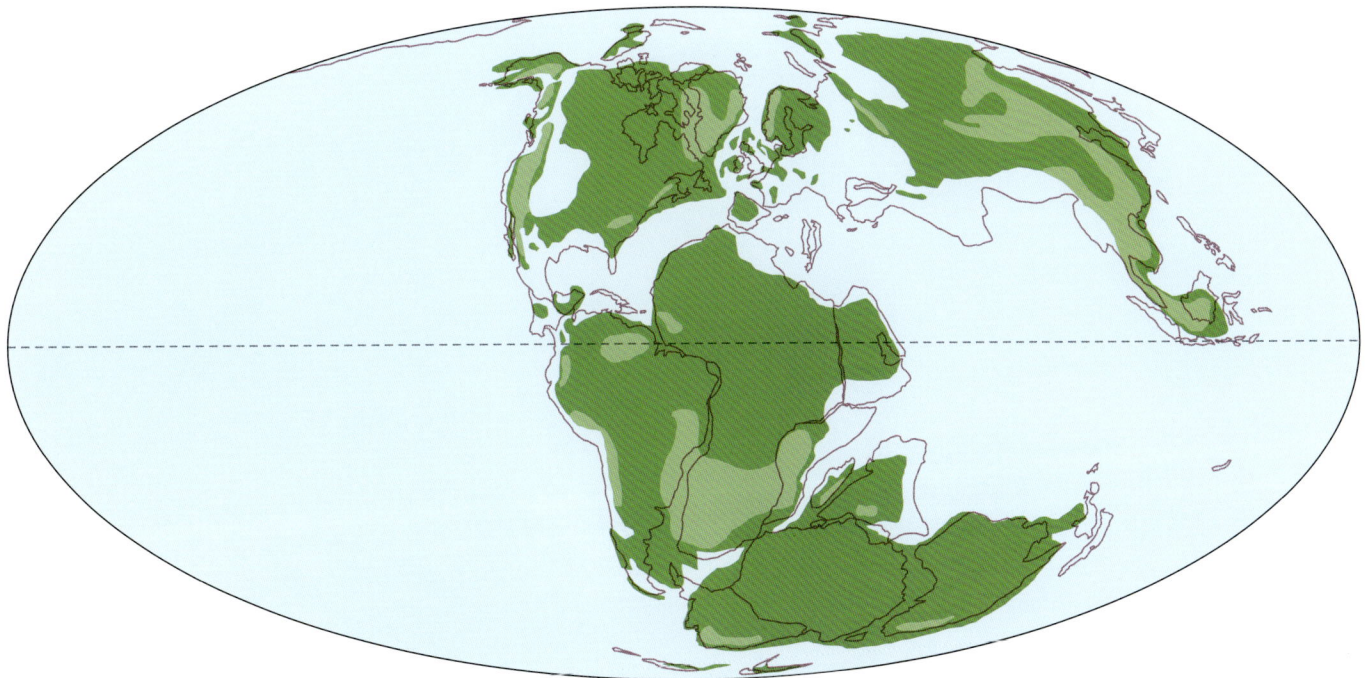

Late Jurassic (Kimmeridgian)

Early Cretaceous (Valanginian–Berriasian)

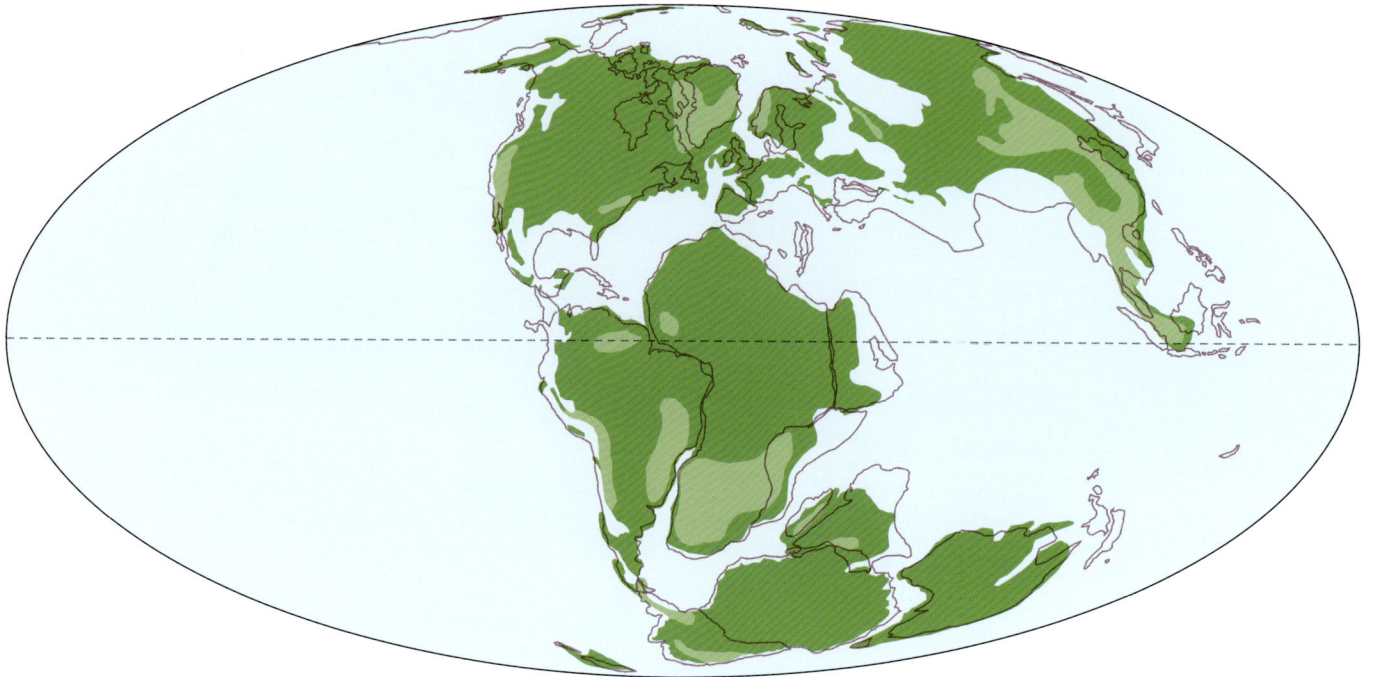

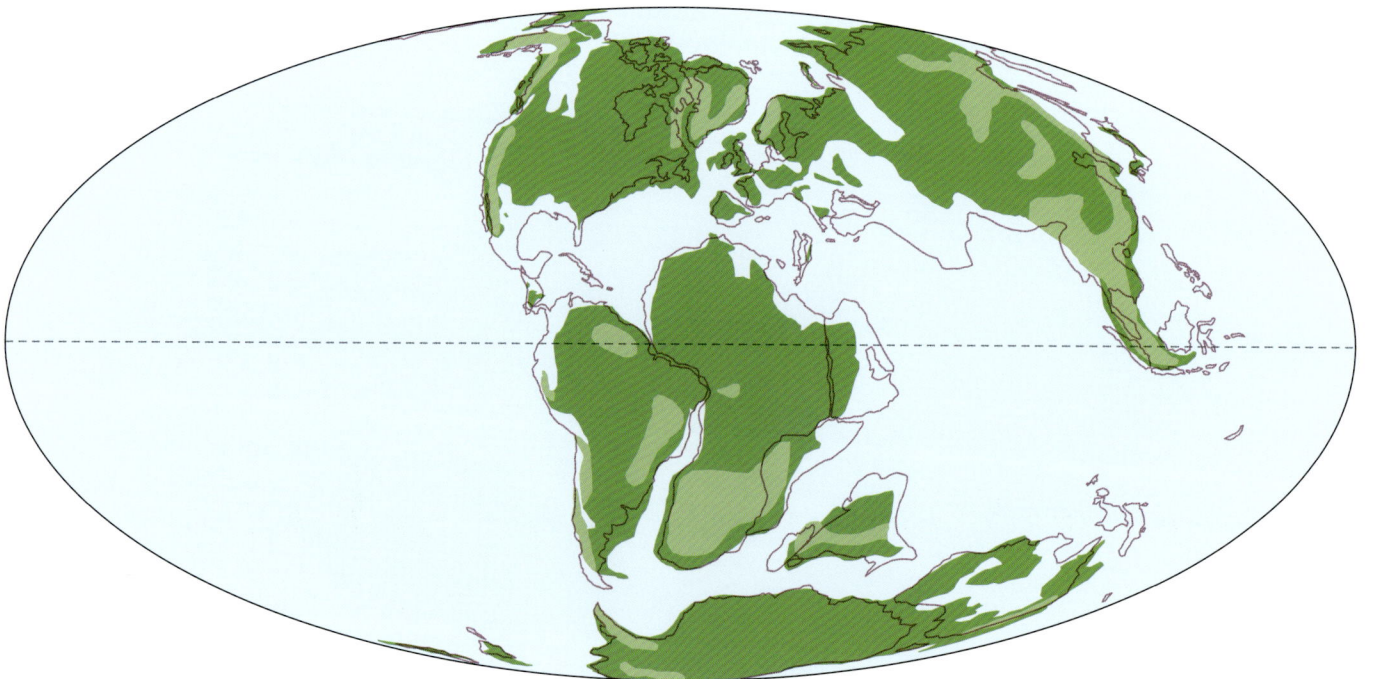

Early Cretaceous (Aptian)

Late Cretaceous (Coniacian)

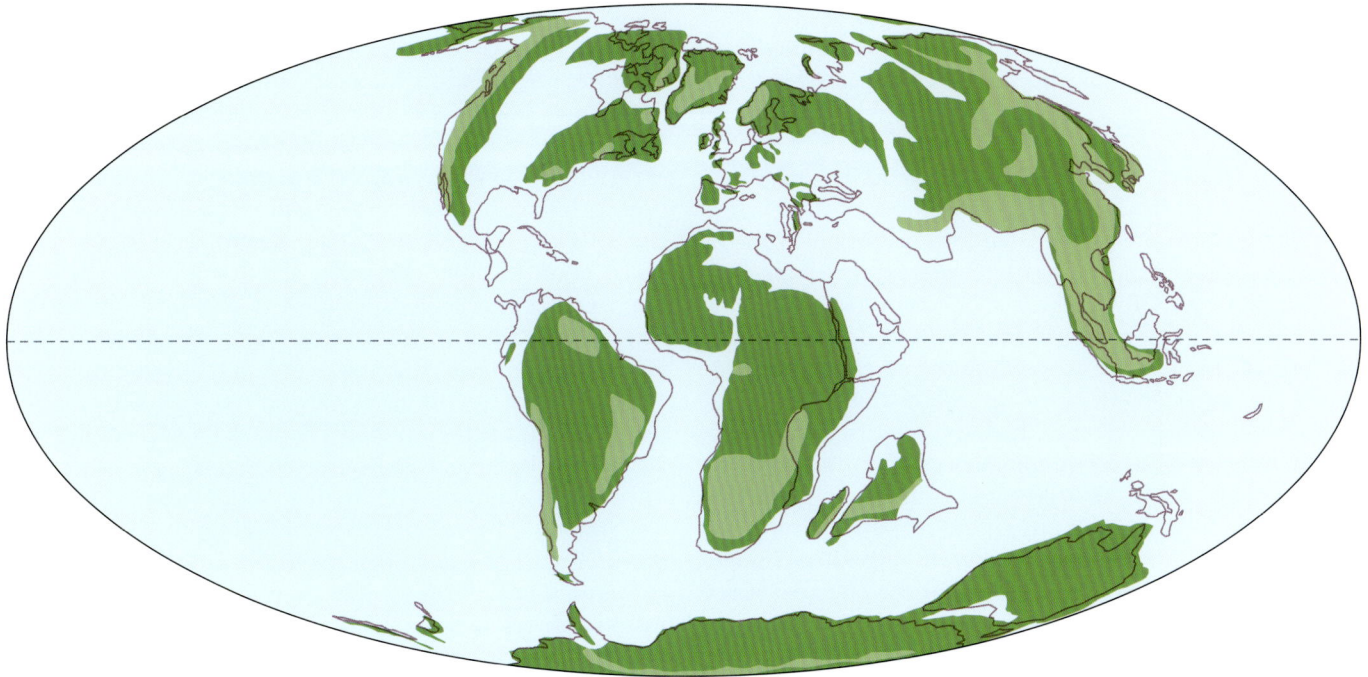

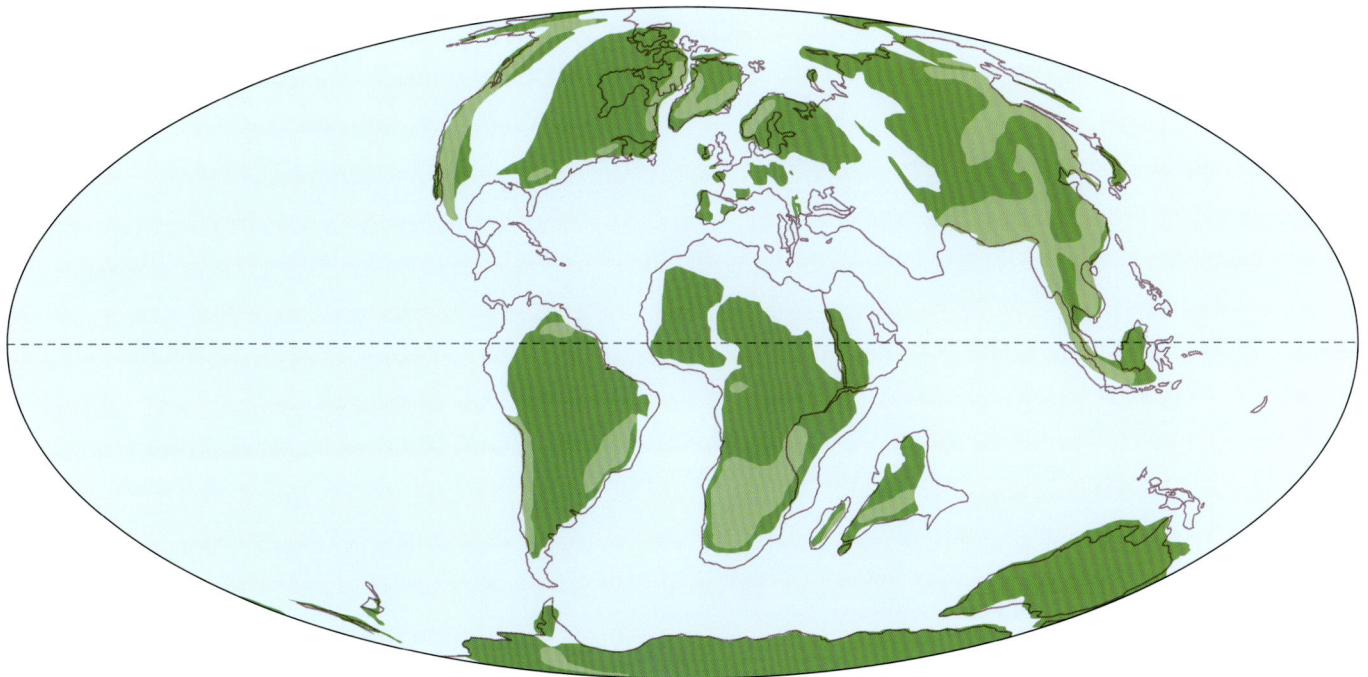

Late Cretaceous (Campanian)

All but a very small percentage of carcasses are destroyed at or soon after death. Many are consumed outright—particularly the juveniles and small adults—or in part by predators and scavengers, and almost all the rest rotted or are weathered away before or after burial at some point. Even so, the number of animals that have lived over time is immense. Because at any given time, a few hundred million sauropodomorphs were probably alive—judging by the populations of living land mammals of similar size—mostly juveniles and small adults, and the groups existed for most of the Mesozoic, the number of their fossils that still exist on the planet is enormous, probably in the many hundreds of millions or billions of individuals.

Of these, only a tiny fraction of 1 percent has been found at or near the small portion of the sauropodomorph-bearing formations that are exposed on the surface, where the fossils can be accessed, or in the mines that allow some additional remains to be reached. Even so, the number of sauropodomorph fossils that have been scientifically documented to at least some degree is considerable. Some sauropodomorph bone beds contain the remains of hundreds of individuals, and the total number of individuals known in that sense is probably in the thousands. The question is where to find them.

Much of the surface of the planet at any given time is undergoing erosion. This is especially true of highlands. In erosional areas, sediments that could preserve the bones and other traces are not laid down, so highland faunas are rarely found in the geological record, the Yixian–Jiufotang lake-forest district being a notable exception, deposited in volcanic uplands. Fossilization has the potential to occur in areas in which sediments are being deposited quickly enough, and in large enough quantities, to bury animal remains and traces before they are destroyed by consumption or exposure. Animals can be preserved in deep fissures or caves in highland areas; this is fairly rare but not unknown when dealing with the Mesozoic. Areas undergoing deposition tend to be lowlands downstream of uplifting or volcanic highlands that provide abundant sediment loads carried in streams, rivers, lakes, or lagoons that settle out to form beds of silt, sand, or gravel. Therefore, large-scale formation of fossils occurs only in regions experiencing major tectonic and/or volcanic activity. Depositional lowlands can be broad valleys or large basins of varying size in the midst of highlands, or coastal regions. As a result, most known sauropodomorph habitats were flatlands, usually riverine floodplains, with little in the way of local topography. In some cases, the eroding neighboring highlands were visible in the distance from the locations where fossilization was occurring; this was especially true in ancient rift valleys and along the margins of large basins. Ashfalls can preserve skeletons en masse, but lava flows tend to incinerate and destroy animal remains. In deserts, windblown dunes preserve bones, and also when they slump when wetted by rains. Also suitable for preserving the occasional sauropod carcass as floating drift are sea and ocean bottoms.

Most sediment deposition occurs during floods, which may also drown animals that are then buried and preserved. The great majority of preserved dinosaurs, however, died before a given flood. The bottoms and edges of bodies of water, whether streams, rivers, or lakes, are prime preservation locations. In some cases, these watery locations lead to exquisite preservation, including soft tissues. Also good are floodplains, although nonburial before the next seasonal flood results in their degradation by feeding vertebrates and invertebrates and exposure. An extreme means of fossilization with nearly perfect preservation is tree sap that fossilizes; this can capture only small remains, or animal parts. Once burial occurs, the processes that preserve remains are complex and in many regards poorly understood. It is being realized, for instance, that bacterial activity is often important in preserving organic bodies. Depending on the circumstances, fossilization can be rapid or very slow, to the point that it never really occurs even after millions of years. The degree of fossilization therefore varies and tends to be more extensive the further back in time the animal was buried. The most extreme fossilization occurs when the original bone is completely replaced by groundwater-borne minerals. Some Australian dinosaur bones have, for instance, been opalized. Most dinosaur bones, however, retain the original calcium structure. The pores have been filled with minerals, converting the bones into rocks much heavier than the living bones. In some locations, such as the Morrison Formation, bacterial activity encouraged the concentration of uranium in many bones—skeletons still below the surface can be detected by their radiation—leading to a radiation risk from stored bones. In other cases, the environment surrounding dinosaur bones has been so stable that little alteration has occurred, leading to the partial retention of some soft tissues near the core of the bones.

While sauropodomorph bones are on the scarce side, their trackways are much more abundant, what with a single individual able to leave behind lots of prints—an active animal can make tens of thousands of steps in a day, a few million a year, a hundred million or so in a lifetime. Prints are left in sediments sufficiently moist to record them, usually along the edges of bodies of water from streams, rivers, ponds, lakes, and ocean beaches. Sometimes they are stabilized by algal and bacterial mats that keep them from weathering away until the next influx of sediment preserves them. Or windblown sediments are the initial agent of preservation. Or the prints may be degraded, especially if made in soft muds that quickly slumped. Wet places are not necessary to retain prints; those of some prosauropods have been found on the flanks of fossil dunes. Most evocative are tracks associated with raindrop impressions from showers too brief to ruin the prints. Tracks are usually revealed by horizontal splitting of slabs. In that case, it is not always true that the print on view is the one directly impressed by the foot; if the print lacks foot-pad scales, then the split layer may or may not be one above or below the surface stepped upon back in the Mesozoic. Sometimes prints are exposed in a vertical section as profiles. Generally, where trackways are common, bones are not, although this separation may be between levels in the same formation. Some places are dinosaur-trackway freeways dense with their trails, from little species and juveniles to giant adults; these are usually shoreline deposits that sauropodomorphs were using as open traffic corridors. As for other trace fossils, don't forget the poop—coprolites—which can be fairly abundant in some places. Also, polished gastroliths may be present scattered in sediments that are otherwise barren of rocks. Eggs, often at least their broken shells, and nests are not extremely rare; having been buried by the parents, their preservation is not all that surprising.

Although the number of prosauropod and sauropod bones and trackways that lie in the ground is tremendous, all but a tiny fraction are for practical reasons out of reach. Nearly all are simply buried too deeply. The great majority of fossils that are found are on or within a few feet of the surface. Even if deposits loaded with sauropodomorph fossils occur near the surface, their discovery is difficult if a heavy cover of well-watered vegetation and soil hides the sediments. For example, large tracts of dinosaur-bearing Mesozoic sediments lie on the Eastern Seaboard, running under major cities such as Washington, DC, and Baltimore. But the limited access to the sediments hinders discoveries, including sauropods, which are largely limited to construction sites made available by willing landowners.

Prime sauropodomorph real estate consists of suitable Mesozoic sediments that have been exposed and eroded over large areas that are too arid to support heavy vegetation. This includes short-grass prairies, badlands, and deserts. In some locations, countless trackways have been exposed. In most cases, dinosaur bones are much less common. Finding sauropodomorphs has changed little since the 1800s. It normally consists of walking slowly, stooped over, usually under a baking sun, often afflicted by flying insects, sometimes by the venomous snake, looking for telltale traces. If really small remains are being looked for, such as fragmented eggshells, crawling on (padded) knees is necessary. Novices often miss the traces against the background of sediments, but even amateurs soon learn to mentally key in on the characteristics that distinguish fossil remains from rocks. Typically, broken pieces of bones on the surface indicate that a bone or skeleton is eroding out. I once missed finding a giant sauropod skeleton because I wanted to get back to camp for dinner. Another expedition member went up a side gully and found bone fragments. One hopes that tracing the broken pieces upslope will soon lead to bones that are still in place. In recent years, GPS has greatly aided in determining and mapping the position of fossils. Ground-penetrating radar has sometimes been used to map out the extent of a newly found set of remains, but usually researchers just dig and see what turns up.

Now it becomes a matter of properly excavating and removing the fossil without damaging it while scientifically investigating and recording the nature of the surrounding sediments in order to recover the information they may contain. These basic methods have also not changed much over the years. On occasion, thick overburden may be removed by heavy equipment or even explosives. But usually it is a job for jackhammers, sledgehammers, picks, or shovels, depending on the depth and hardness of the sediments and the equipment that can be brought in. When the remains to be recovered are reached, more-careful excavation tools, including trowels, hammers and chisels, picks, and even dental tools and brushes, are used. It is rare to be able to simply brush sand off a well-preserved fossil as in the movies, although this happy circumstance does occur in some ancient dune deposits in Mongolia. Usually, sediments are cemented to some degree and require more forceful action. At the same time, the bones and other remains are fragile, and care must be taken to avoid damaging them. And their position has to be documented by quarry maps, photography, or laser scanning before removal. Individual bones can be removed, or blocks of sediment including multiple bones or articulated skeletons may have to be taken out intact. Again, these operations are usually conducted under conditions that include flying insects, dust, heat, and sun, although tarps can provide shade.

After exposure, especially fragile bones may be soaked with glue to harden them. On the other hand, the increasingly sophisticated techniques being applied to bones in the laboratory discourage alteration and contamination of bones. Before removal, most remains are quickly covered with tissue paper that is wetted in place, followed by heavier paper, and over that a thick layer of plaster to form a protective jacket. Wood is usually used to brace the jacket. When the top is so protected, the remains are undermined and then flipped—a process often requiring considerable exertion and entailing some risk to both excavators and the fossils. Then the other side is papered and plastered, forming a protective cocoon. These techniques have changed little in over century. If the jacketed block is very heavy and not accessible to heavy equipment, a helicopter may be brought in to lift it out. On occasion, this requires a heavy-lift helicopter; the US Army is sometimes willing to conduct such operations gratis as part of dissimilar-cargo training that allows its crews to learn how to cope with challenging objects, rather than standard pallets.

Because dinosaur paleontology is not a high-priority science backed by large financial budgets, and because the number of persons searching for and excavating dinosaurs in the world in a given year is only a few thousand—albeit far more than in the past—the number of sauropodomorph skeletons that now reside in museums is still just a few thousand.

In the lab, preparators remove part or all of the jacket and use fine tools to eliminate some or all of the sediment from the bones and any other remains. Most bones are left intact, and only their surface form is documented. In some situations, chemical treatment is required to stabilize bones; this is especially true if the bones are impregnated with pyrite, which gradually swells with moisture. Increasingly, certain bones are opened to reveal their internal structure for various purposes: sectioning is used to examine bone histology and microstructure, to count growth rings, to search for traces of soft tissues, and to sample bone isotopes and proteins. It is becoming the norm to conduct CT scans on skulls and complex bones as a means to determine the three-dimensional structure without invasive preparation, as well as to reduce costs. These can be published as conventional hard copies and in digital form. There is increasing reluctance to put original bones in mounted skeletons in display halls because delicate fossils are better conserved when properly curated in the back areas. And mounting heavy, fragile bones poses many difficulties and limitations while removing them from ready study. Instead, the bones are molded, and lightweight casts are used for the display skeleton.

There has never been as much sauropodomorph-related activity as there is today. At the same time, there is the usual shortage of funding and personnel compared to what could otherwise be done. The happy outcome is that there are plenty of opportunities for amateurs to participate in finding and preparing dinosaurs. If nonprofessionals search for fossils on their own, they need to pay attention to laws and to paleontological ethics. In some countries, all sauropodomorph fossils are regulated by the state—this is true in Canada, for instance. In the United States, fossils found on privately held land are entirely the property of the landowner, who can dispose of any prehistoric remains as he or she sees fit. Any search for and retention of fossils on private property is therefore by permission of the owner, many but not all of whom are interested in the fossils on their land. Because dinosaur remains in the eastern states usually consist of teeth and other small items, nonhazardous construction sites are often available for exploration on non-workdays. In the West, ranches with cooperative owners are primary sources of dinosaur remains. The rising sums of money to be made by selling fossils are making it more difficult for scientific teams to access such lands, but private collectors reveal specimens that would otherwise not be known. The religious opinions of some landowners are also an occasional barrier. Fortunately, dinosaur fossils are a part of western lore and heritage, so many locals favor paleontological activities, which contribute to the tourist trade. All fossils on federal government land are public property and are heavily regulated. Removal can occur only with official permission, which is limited to accredited researchers. Environmental concerns may be involved because dinosaur excavations are in effect small-scale mining operations. Fossils within Native American reservations may likewise be regulated, and collaboration with Original Peoples is indispensable. In any case, deep cuts in government spending on science, including paleontological, threaten to seriously curb the discovery of and research on dinosaur fossils. Sauropodomorph fossils found by nonprofessionals searching on their own should not be disturbed. Instead, they should be reported to qualified experts, who can then properly document and handle the remains. In such cases, the professionals are glad to do so with the assistance of the discoverer.

A number of museums and other institutions offer courses to the public on finding, excavating, and preparing sauropodomorphs and other fossils. Most expeditions include unpaid volunteers who are trained, often on-site, to provide hands-on assistance to the researchers. Participants are usually expected to pay for their own transportation and general expenses, although food and in some cases camping gear as well as equipment may be provided. In order to tap into the growing number of dinosaur enthusiasts, commercial operations led by qualified experts provide a sauropodomorph-hunting experience for a fee, usually in the western states and Canada. Those searching for and digging up dinosaurs need to take due precautions to protect themselves from sunlight and heat, in terms of UV exposure, dehydration, and hyperthermia, as well as biting and stinging insects and scorpions. Rattlesnakes are often common in the vicinity of dinosaur fossils. Steep slopes, cliffs, and hidden cavities are potential dangers. In many sauropodomorph formations, gravel-like caliche deposits that formed in the ancient, semiarid soils create roller-bearing-like surfaces that undermine footing. Flash floods can hit quarries or ill-placed campsites. The use of mechanical and handheld tools when excavating fossils poses risks, as does falling debris from quarry walls. When impact tools are used on hard rock, eye protection may be necessary. Chemicals used while working with fossils require proper handling.

Back in the museums and other facilities, volunteers can be found helping prepare fossils for research and display, and cataloging and handling collections. This is important work because, in addition to the constant influx of new fossils with each year's harvest, many dinosaur fossils found as long as a century ago have been sitting on shelves, sometimes still in their original jackets, without being researched.

Landowners who allow researchers onto their land sometimes get a new species found on their property named after them, informally or formally. So do volunteers who find new sauropodomorphs. Who knows? You may be the next lucky amateur.

USING THE GROUP AND SPECIES DESCRIPTIONS

Over 300 sauropodomorph species have been named, but a substantial number of these names are invalid. Many are based on inadequate remains, such as teeth or one or a few bones, that are taxonomically indeterminate. Others are junior synonyms for species that have already been named. *Seismosaurus*, for instance, proved to be the same as the previously named *Diplodocus*, so the former is no longer used. This guide includes mainly those species that are generally considered valid and are based on sufficient remains. A few exceptions are allowed when a species based on a single bone or little more is important in indicating the existence of a distinctive type or group of dinosaurs in a certain time and place. Many of the group and species entries have been changed very little if at all from the first to third editions of the general dinosaur guides; corrections and new information have caused a substantial minority to be more heavily revised and, in a few cases, dropped, at the same time that new species have been added.

The species descriptions are listed hierarchically, starting with major groups and working down the ranking levels to genera and species. Because many researchers have abandoned the traditional Linnaean system of classes, orders, suborders, and families, there is no longer a standard arrangement for the dinosaurs—many sauropodomorph genera are no longer placed in official families—so none is used here. In general, the taxa are arranged phylogenetically. This presents a number of problems. It is more difficult for the general reader to follow the various groupings. Although there is considerable consensus concerning the broader relationships of the major groups, at lower levels the incompleteness of the fossil record hinders a better understanding. The great majority of sauropodomorph species are not known, many of those that are known are documented by incomplete remains, and it is not possible to examine dinosaur relationships with genetic analysis. Because different cladistic analyses often differ substantially from one another, I have used a degree of personal choice and judgment to arrange the groups and species within the groups. Some of these placements reflect my considered opinion, while others are arbitrary choices between competing research results. Most of the phylogeny and taxonomy offered here is not a formal proposal, but a few new group labels were found necessary and are coined and defined here for future use by others if it proves efficacious. Disputes and alternatives concerning the placement of dinosaur groups and species are often but not always mentioned.

Under the listing for each prosauropod and sauropod group, the overall geographic distribution and geological time span of its members are noted. This is followed by the anatomical characteristics that apply to the group in general, which are not repeated for each species in the group. The anatomical features usually center on what is recorded in the bones, but other body parts are covered when they have been preserved. The anatomical details are for purposes of general characterization and identification and reflect as much as possible what a dinosaur watcher might use; they are not technical phylogenetic diagnoses. The type of habitat that the group favored is briefly listed, and this varies from specific in some types to very generalized in others. Also outlined are the restored habits that probably characterize the group as a whole. The reliability of these conclusions varies greatly. There is, for example, no doubt that sauropodomorphs with blunt teeth consumed far more vegetation than flesh. Nor is there good reason to

doubt that long-necked sauropods fed high in tree canopies. Less certain is how bipedal versus quadrupedal prosauropods with medium-length arms were.

The naming of sauropodomorph genera and species is often problematic, in part because of a lack of consistency. It has been recognized of late that the apatosaurs were overlumped, resulting in the revival of *Brontosaurus*, while brachiosaurs were oversplit, leading to the creation of *Giraffatitan*. In this work, I have attempted to apply more uniform standards to generic and species designations, with the divergences from the conclusions of others noted.

Also a big problem is the inadequate nature of the fossils very many dinosaur taxa are based upon. Such is true of the original fossils of *Apatosaurus*, *Diplodocus*, *Brachiosaurus*, *Plateosaurus*, *Argentinosaurus*, and *Alamosaurus*. But sorting out species, genera, and families is often seriously hindered; these items are noted.

The entry for each species first cites the dimensions and estimated mass of the taxon. The total length is for the combined skull and skeleton along the line of the vertebrae; the degree of completeness of the skull plus vertebral series impacts the accuracy of the given length. The values are general figures for the size of the largest known adults of the species and do not necessarily apply to the values estimated for specific fossils, most of which can be found via https://press.princeton.edu/books/hardcover/9780691268651/the-princeton-field-guide-to-prosauropod-and-sauropod-dinosaurs, which includes the mass estimates for each included fossil. Because the number of fossils for a particular species is a small fraction of those that lived, the largest individuals are not measured; "world record" fossils can be a third or more heavier than is typical—recent estimates that the largest theropod species got to be two-thirds larger are improbable. The sizes of species known only from immature fossils are not estimated. All values are, of course, approximate, and their quality varies depending on the completeness of the remains for a given species. If the species is known from sufficiently complete remains, the dimensions and mass are based on the profile-skeletal restoration. These are restored with the animal in normal lean-healthy condition, not carrying heavy fat deposits as can occur among some herbivores by the end of the optimal feeding season of the year—also avoided is the extreme shrink-wrapped look of malnourished animals that has been seen in some restorations. The latter is used to estimate the volume of the dinosaur, which can then be used to calculate the mass, with the portion of the volume that was occupied by lungs and any air sacs taken into account. It has been realized of late that animals are generally denser than had been thought. Most land mammals when swimming, for example, are nearly awash and avoiding drowning only by swimming, so the density, or specific gravity (SG), is set at 1.0 relative to the density of water for prosauropods having little or nothing in the way of air sacs. Armored dinosaurs were likely significantly denser than water, so their SGs are 1.06–1.1. The big ratites, whose leg bones are hollow, float a little higher in water than mammals, suggesting SGs of 0.92. Because pneumatic theropods and sauropods lacked airy leg bones, their SG was probably between that and 0.96. A complication with sauropods is that their necks would have been less dense than the rest of the animal, and neck volume varied a lot relative to the latter; modeling indicates the SG of their necks was 0.85–0.9, calculated separately. Because these plant eaters are drawn with fodder-filled, rotund bellies, the mass estimates include this large amount of digesting vegetation, this being the normal situation for healthy, well-fed herbivores. When remains are too incomplete to directly estimate dimensions and mass, masses are extrapolated from those of relatives and are considerably more approximate. Both metric and English measurements are included except for metric tonnes, which equal 1.1 English tons; all original calculations are metric, but because they are often imprecise, the conversions from metric to English are often rounded off as well.

The next line outlines the fossil remains, whether they are skull or skeletal material or both, that can be confidently assigned to the species to date; the number of fossils varies from one to thousands. The accuracy of the list ranges from exact to a generalization. The latter sometimes results from recent reassignment of fossils from one species to another, leaving the precise inventory uncertain. Skeletal and/or skull restorations have been rendered for those species that are known from sufficiently complete remains that have been adequately documented by the time the book was being finalized to execute a reconstruction—or for species that are of such interest that a seriously incomplete restoration is justified. A number of species known from good remains have yet to be made available for research, sometimes decades after their discovery. In some cases, only oblique-view photographs unsuitable for a restoration of reasonably complete skulls and skeletons are obtainable. The pace of discovery is so fast that a few new finds could not be included. Despite the absences, this is by far the most extensive sauropodomorph skull-skeletal library yet published in print. The core skeletal fossils that have been restored can be found at the above link. The restorations show the bones as solid white set within the solid black restored muscle and keratin profiles; cartilage is not included. Absence of most or all of the skull often precludes a restoration, those dinosaurs with small skulls being the most common exception. In many cases, only the skull is complete enough for illustration. When more than one skeleton and/or skull of a given species is included, they are always to the same scale to facilitate size comparisons. In most cases, the skull and skeletal restorations are of adults or close to it, but growth series are included for all examples that are achievable, on the premise that young dinosaurs soon left their nests and were part of the general fauna. In some examples, the skeletons of the smallest juveniles are reproduced both to scale to the much larger adults, and at a larger scale to allow details to be seen better—this guide contains the largest set to date of dinosaur growth stages. The anatomical and proportional accuracy of the restorations ranges from very exacting for those that are known from extensive remains that have been described and measured in detail and/or have good photographs on hand—those who wish to criticize the accuracy of these meticulous illustrations need to provide measurements that indicate otherwise—down to approximate if much of the species remains are missing or have not been well illustrated. The restorations have been prepared over four decades to fulfill differing requirements. A number of skeletons and skulls show only those bones that are known—sometimes with replications of one side to the other, ranging from a large fraction to nearly all—whereas others have been filled out to represent a complete skeleton. Reliable information about exactly which bones have and have not been preserved is often not available, so the widely used term "rigorous restoration" of incomplete profile-skeletals is best avoided in favor of "known bone." Top views of skeletons have been provided when such were doable, as well as fore and aft views in a number of cases. The skeletons are posed in a common, basic posture, with the right hindleg pushing off at the end of its propulsive stroke, in order to facilitate cross comparisons. The prosauropods are shown in a trot when quadrupedal, the sauropods in an amble. The necks and tails are also not posed in a strictly neutral osteological posture; more dynamic curvatures are utilized—the very slender tail whips are shown drooping from the muscular portion of the tail as they probably did in life. Representative examples of shaded skull restorations have been included with some of the major groups. The same has been done with a sample of muscle studies, whose detailed nature is no less or more realistic than are the particulars in the full-life restorations, which, if anything, involve additional layers of speculation.

The color plates are based on species for which the fully or nearly adult skeletal restorations or skulls are deemed of sufficient quality for a full-life restoration. If it is unlikely that new information will significantly alter the skeletal plan in the near future, a life restoration has been prepared; if the skeletal is not sufficiently reliable, a life restoration has not been done. A few of the earlier life restorations have been replaced because of updates of their skeletal restorations. Most of the skull as well as the skeleton needs to be present to justify a life restoration of an entire animal. In a few cases, only the skull is good enough to warrant a life restoration to the exclusion of the overall body. The colors and patterns are speculative for sauropodomorphs, there so far being no examples where coloration has been restored based on preserved skin pigmentation. Extremely vibrant color patterns have generally not been used to avoid giving the impression that they are identifying features. Those who wish to use the skeletal, muscle, and life restorations herein as the basis for commercial and other public projects are reminded to first contact the copyright illustrator.

The particular anatomical characteristics that distinguish the species are listed, but these too are for purposes of general identification by putative dinosaur watchers, not for technical species diagnoses. These differ in extent depending on the degree of uniformity versus diversity in a given group, as well as the completeness of the available fossil remains. In some cases, the features of the species are not different enough from those of the group to warrant additional description. In other cases, not enough is known to make a separate description possible.

Listed next is the formal geological time period and, when available, the specific stage that the species is known from, and the segment of that stage, from early to middle to late. As discussed earlier, the age of a given species is known with a precision of within a million years in some cases, or as poorly as an entire period in others. The reader can refer to the timescale on the timeline chart to determine the age, or age range, of the species in years (see pp.

84–85). Because most species exist for only a few hundred thousand years before either being replaced by a descendant species or going entirely extinct, when fossils within a single genus are found in sediments that span more than that amount of time, it is presumed they are different species. In some cases, it is not entirely clear whether a species was present in just one time stage or crossed the boundary into the next one. In those cases. the listing includes "and/or," such as late Santonian and/or early Campanian.

Next, the geographic location and geological formation from which the species is so far known are listed. The paleomaps of coastlines (see pp. 74–77) can be used to geographically place a species in a world of drifting continents and fast-shifting seaways, with the proviso that no set of maps is extensive enough to show the exact configuration of the ancient lands when each species was extant. Some dinosaur species are known from only a single location and formation, whereas others have been found in an area spread over one or more formations. In the latter situation, the first location/formation listed is for the original fossil. I have tended to be conservative in listing the presence of a specific species only in those places and levels where sufficiently complete remains are present. Some formations have not been named, even in areas that are well studied. Many formations were formed over a time span that was longer than that of some or all of the species that lived within them, so, when possible, the common procedure of simply listing just the overall formation a given species is from is avoided. Because the Morrison

Riojasaurus, Riojasuchus, and *Pseudhesperosuchus*

83

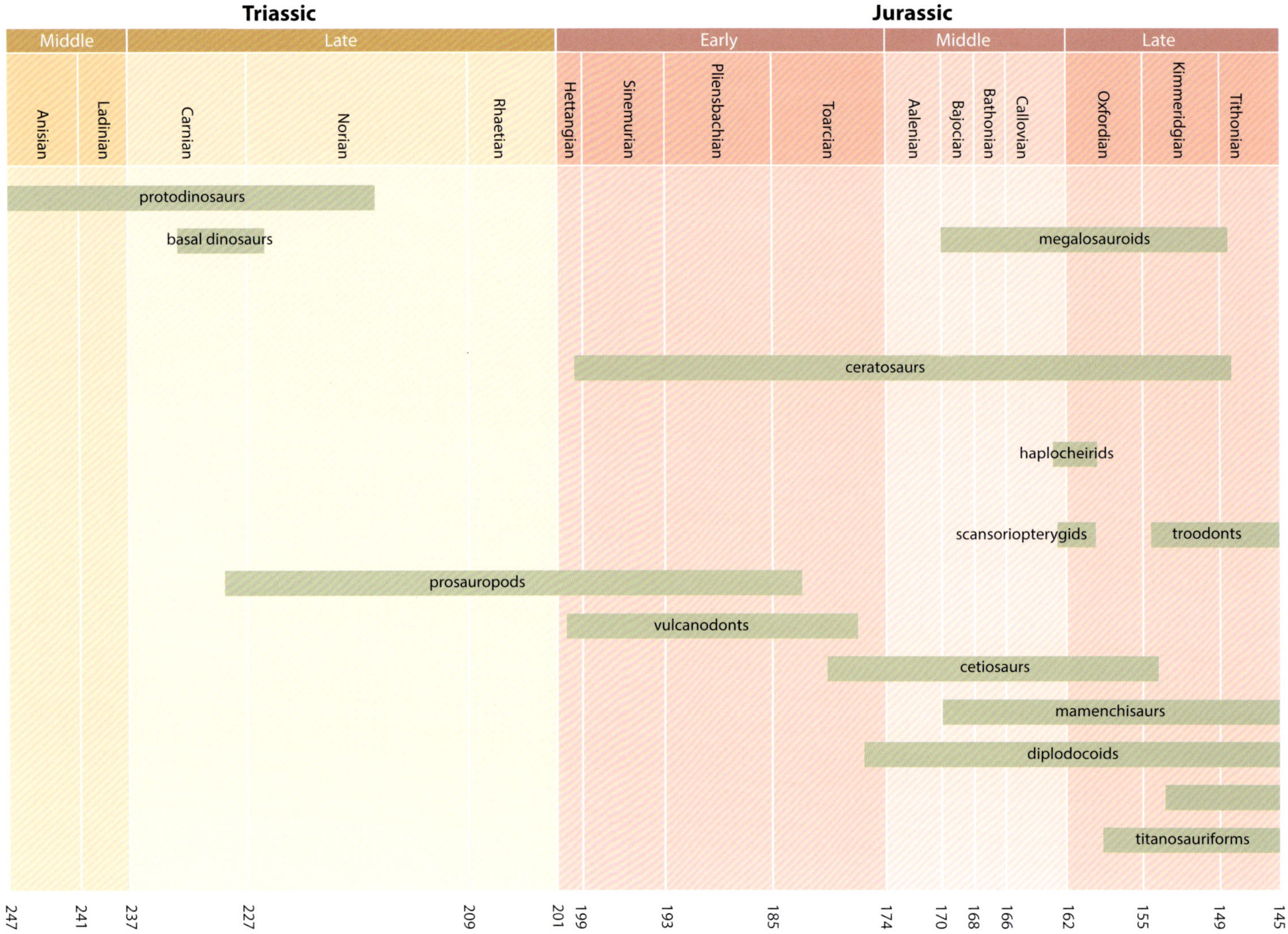

	Triassic			Jurassic		
Middle	Late			Early	Middle	Late

Anisian · Ladinian · Carnian · Norian · Rhaetian · Hettangian · Sinemurian · Pliensbachian · Toarcian · Aalenian · Bajocian · Bathonian · Callovian · Oxfordian · Kimmeridgian · Tithonian

protodinosaurs
basal dinosaurs
megalosauroids
ceratosaurs
haplocheirids
scansoriopterygids
troodonts
prosauropods
vulcanodonts
cetiosaurs
mamenchisaurs
diplodocoids
titanosauriforms

247 · 241 · 237 · 227 · 209 · 201 · 199 · 193 · 185 · 174 · 170 · 168 · 166 · 162 · 155 · 149 · 145

Formation was deposited over a span of six million years in the Late Jurassic, there was extensive change over time in the apatosaurines, diplodocines, camarasaurs, and haplocanthosaurs that dwelled in the area. Because the overlying Cedar Mountain Formation formed over 40 million years—suggesting a need to split the oversized formation into smaller formations—the sauropod fauna underwent radical changes. I have therefore listed the level of the formation that each species comes from, when this information is available. The reader can get an impression of what dinosaur species constituted a given fauna in a particular bed of sediments by using the formation index. Only species present at a given level of a formation could have lived together, but this does not always mean a given set of dinosaurs listed as being from a particular level did so, because time separations sometimes exist. In a few cases, the

sediments a dinosaur is from are not yet named—often but not always because they have not been studied—and the geological group may be named instead, if one is available.

Noted next are the basic characteristics of the dinosaur's habitat in terms of rainfall and vegetation, as well as temperature when it is not generally tropical or subtropical year-round. Environmental information ranges from well studied in heavily researched formations to nonexistent in others. If the habits of the species are thought to include attributes not seen in the group as a whole, then they are outlined. Listed last are special notes about the species when they are called for. Possible ancestor-descendant relationships with close older or younger relatives are sometimes noted, but these are always tentative. This section is also used to note alternative hypotheses and controversies that apply to the groups and species.

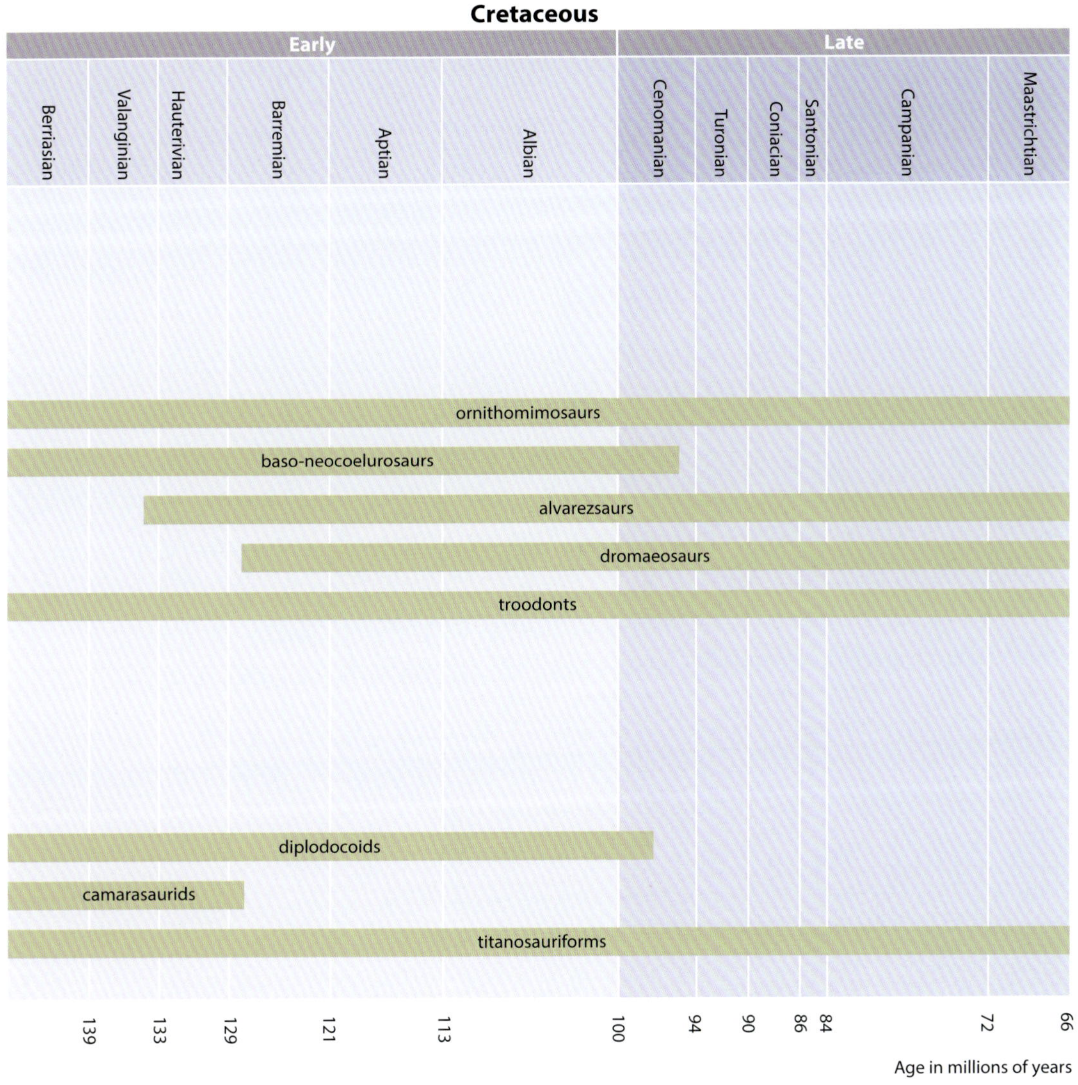

Cretaceous

	Early						Late					
	Berriasian	Valanginian	Hauterivian	Barremian	Aptian	Albian	Cenomanian	Turonian	Coniacian	Santonian	Campanian	Maastrichtian

ornithomimosaurs

baso-neocoelurosaurs

alvarezsaurs

dromaeosaurs

troodonts

diplodocoids

camarasaurids

titanosauriforms

139 133 129 121 113 100 94 90 86 84 72 66

Age in millions of years

Giraffatitan brancai

SAUROPODOMORPHS

SMALL TO COLOSSAL HERBIVOROUS AND OMNIVOROUS SAURISCHIAN DINOSAURS FROM THE LATE TRIASSIC TO THE END OF THE DINOSAUR ERA, ALL CONTINENTS.

ANATOMICAL CHARACTERISTICS Moderately variable. Heads small, not strongly constructed, nostrils enlarged, teeth blunt, nonserrated. Necks not stout, 10–19 cervicals. Tails long. Arms and legs neither elongated nor slender. Five fingers. Pelves small to large, five to four toes. Largely bipedal to always quadrupedal when moving, all able to rear up on hindlegs. Brains reptilian. Gizzard stones sometimes present, used either to help grind or to stir up ingested food. In trackways, hands usually farther from midline than feet, never closer, able to rear up with varying degrees of ease.

HABITATS Very variable, deserts to well-watered forests, tropics to polar regions.

ENERGETICS Thermophysiology intermediate to very high energy.

REPRODUCTION AND ONTOGENY Usually reached sexual maturity while still growing, most or all rapid breeders. Many laid hard-shelled eggs in pairs, others may have laid soft-shelled eggs. Growth rates moderate to rapid.

HABITS Predominantly herbivorous browsers and grazers, although would have been prone to pick up and consume small animals, did not extensively chew food before swallowing. Main defense clawed feet and tails. Adept surface swimmers, so able to cross large bodies of water.

PROSAUROPODS

SMALL TO LARGE HERBIVOROUS AND OMNIVOROUS SAUROPODOMORPHS LIMITED TO THE LATE TRIASSIC AND EARLY JURASSIC, ALL CONTINENTS.

ANATOMICAL CHARACTERISTICS Fairly uniform. Heads lightly built, top of snouts fairly broad across. 10 cervicals. Tails moderately long. Skeletons not pneumatic, respiratory system poorly understood, except that birdlike system not present. Shoulder girdles not large. Hand short and broad. Grasping fingers fairly long, large claws usually present on most fingers, especially thumb. Pelves short, pubes strongly procumbent. Lower legs about as long as upper, foot fairly long, toes long and flexible, outermost toe very reduced. All able to slow walk quadrupedally, those with long arms mainly quadrupedal, with short arms mainly bipedal, or intermediate. Arms and legs flexed but not elongated or slender, so able to run at modest speeds.

ONTOGENY Growth rates moderate, variable within at least some species.

HABITATS Very variable, deserts to well-watered forests, tropics to poles.

HABITS Some or all may have been omnivores, preferred light vegetation. Main defense standing and lashing out with clawed hands and feet. Small prosauropods may have used clawed hands to dig burrows. May have had some diving ability.

ENERGETICS Thermophysiology probably intermediate, energy levels and food consumption probably low compared to more-derived dinosaurs.

NOTES Fragmentary Early Jurassic *Glacialisaurus hammeri* demonstrates prosauropods dwelled in Antarctica.

Plateosaurus muscle study

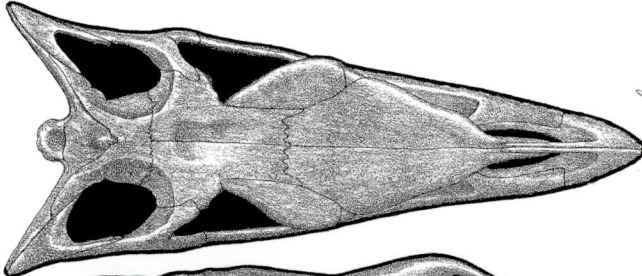

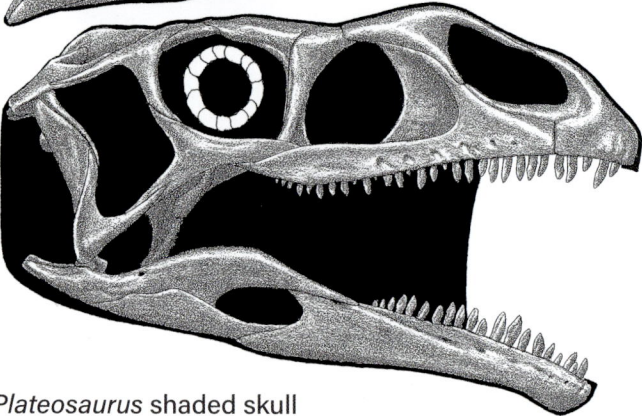

Plateosaurus shaded skull

Whether all the many genera are justified is doubtful. This group is splittable into a number of subdivisions, but relationships within group and with sauropods are uncertain. Many researchers consider known prosauropods to be a sister group to sauropods, but others consider some or all of the first genera to be below the prosauropod-sauropod split, or the latter may have evolved from more-derived prosauropods. Based on supposedly Early Jurassic inadequate remains, Asian *Eshanosaurus deguchiianus* may be a prosauropod, rather than a therizinosaur or other avepod.

BASO-PROSAUROPODS

SMALL PROSAUROPODS LIMITED TO THE LATE TRIASSIC.

ANATOMICAL CHARACTERISTICS Lightly built. Often at least some teeth sharp. Necks neither markedly nor very slender. Thumbs not enlarged.
NOTES Baso-prosauropods and prosauropods excluding eusauropods. Limited distribution may reflect insufficient sampling.

Eoraptor lunensis
—— 1.7 m (5.5 ft) TL, 4.7 kg (10 lb)
FOSSIL REMAINS Two nearly complete skulls and skeletons, almost completely known.
ANATOMICAL CHARACTERISTICS Back teeth bladed and serrated, front teeth more leaf-shaped. Pubis procumbent.

AGE Late Triassic, late Carnian.
DISTRIBUTION AND FORMATION/S Northern Argentina; lower Ischigualasto.
HABITAT Seasonally well-watered forests, including dense stands of giant conifers.

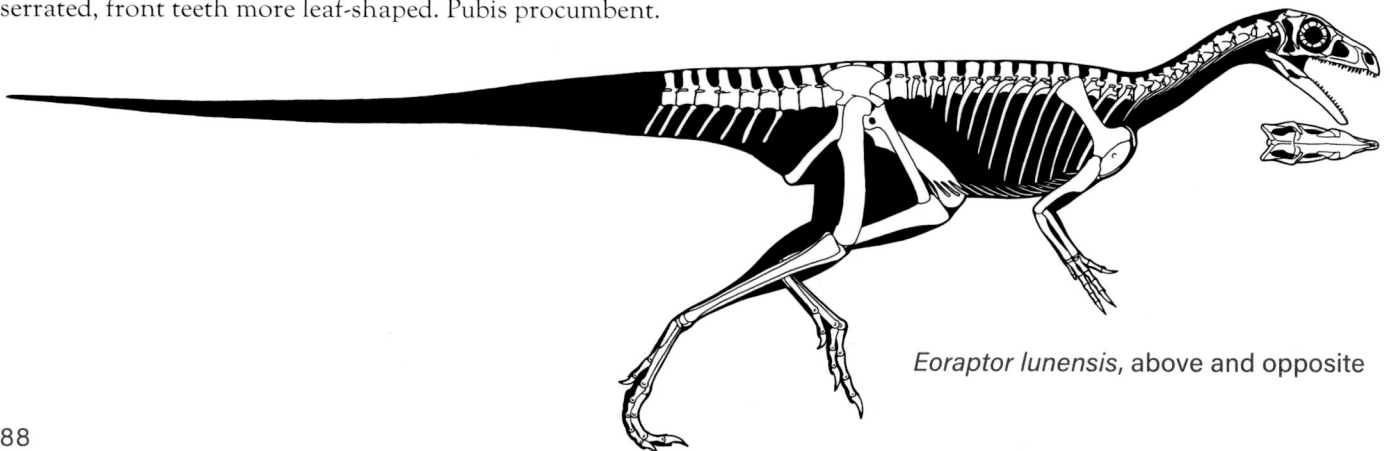

Eoraptor lunensis, above and opposite

HABITS Omnivorous, hunted smaller game and consumed some easily digested plant material.
NOTES Has been considered a very basal saurischian, sauropodomorph, ornithoscelid, and/or theropod, the broad top of snout favors prosauropod status.

Mbiresaurus raathi

2.3 m (7.5 ft) TL, 15 kg (30 lb)
FOSSIL REMAINS Partial skull and majority of skeleton.
ANATOMICAL CHARACTERISTICS Standard for group.
AGE Late Triassic, late? Carnian.
DISTRIBUTION AND FORMATION/S Zimbabwe; Pebbly Arkose.

Guaibasaurus candelariensis

—— 2 m (6.5 ft) TL, 10 kg (20 lb)
FOSSIL REMAINS Partial skeletons.
ANATOMICAL CHARACTERISTICS Insufficient information.
AGE Late Triassic, early Norian.
DISTRIBUTION AND FORMATION/S Southern Brazil; Caturrita.
NOTES Originally thought to be a baso-theropod.

Buriolestes schultzi

—— 1.5 m (5 ft) TL, 4 kg (8 lb)
FOSSIL REMAINS Two skulls and partial skeletons, other remains.
ANATOMICAL CHARACTERISTICS Head shallow, fairly broad, front upper teeth fairly large, most small, most teeth bladed, front lower teeth project somewhat forward. Arm moderately short. Predominantly bipedal.

Eoraptor lunensis

AGE Late Triassic, late Carnian.
DISTRIBUTION AND FORMATION/S Southern Brazil; upper Santa Maria.
HABITS More omnivorous than euprosauropods.
NOTES A somewhat larger partial skeleton from a different location may be this species.

Pampadromaeus barberenai

—— 1.5 m (5 ft) TL, 4.1 kg (8 lb)
FOSSIL REMAINS Majority of skull and skeleton, possible additional partial remains, including juvenile.
ANATOMICAL CHARACTERISTICS Most teeth bladed. Arm moderately short. Predominantly bipedal.
AGE Late Triassic, late Carnian.
DISTRIBUTION AND FORMATION/S Southern Brazil; upper Santa Maria.
HABITS More predaceous than euprosauropods.

Buriolestes schultzi

Pampadromaeus barberenai

Panphagia protos

—— 1.7 m (5.5 ft) TL, 6 kg (12 lb)
FOSSIL REMAINS Partial skull and skeleton.
ANATOMICAL CHARACTERISTICS Front lower teeth bladed. Arm moderately short. Predominantly bipedal.
AGE Late Triassic, late Carnian.
DISTRIBUTION AND FORMATION/S Northern Argentina; lower Ischigualasto.
HABITAT Seasonally well-watered forests, including dense stands of giant conifers.
HABITS More predaceous than euprosauropods.

Saturnalia tupiniquim

—— 1.6 m (5.2 ft) TL, 6 kg (13 lb)
FOSSIL REMAINS Majority of skull and five incomplete skulls and/or skeletons.
ANATOMICAL CHARACTERISTICS Head rather small, snout fairly small. Arm moderately short. Predominantly bipedal.
AGE Late Triassic, late Carnian.
DISTRIBUTION AND FORMATION/S Southern Brazil; upper Santa Maria.
HABITS Small size of head implies pursued very small game.
NOTES *Nhandumirim waldsangae* may be juvenile of this species.

Chromogisaurus novasi

—— 1.7 m (5.5 ft) TL, 5 kg (10 lb)
FOSSIL REMAINS Minority of skeleton.
ANATOMICAL CHARACTERISTICS Insufficient information.
AGE Late Triassic, late Carnian.
DISTRIBUTION AND FORMATION/S Northern Argentina; lower Ischigualasto.
HABITAT Seasonally well-watered forests, including dense stands of giant conifers.

Bagualosaurus agudoensis

—— 1.7 m (5.5 ft) TL, 6 kg (12 lb)
FOSSIL REMAINS Partial skull and skeleton.
ANATOMICAL CHARACTERISTICS Standard for group.
AGE Late Triassic, late Carnian.
DISTRIBUTION AND FORMATION/S Southeastern Brazil; upper Santa Maria.

Thecodontosaurus antiquus

—— 2.5 m (8 ft) TL, 20 kg (40 lb)
FOSSIL REMAINS Majority of skull and partial skeletons, adult to juvenile.
ANATOMICAL CHARACTERISTICS Arm probably moderately short.
AGE Probably Late Triassic, Rhaetian.
DISTRIBUTION AND FORMATION/S Wales; Magnesian Conglomerate?
HABITAT Island.
NOTES *Pantydraco caducus, Asylosaurus yalensis* probably immature examples of this species, may include *Agrosaurus macgillivrayi*, which is probably not from Australia. Found in ancient fissure fills. Name incorrectly implies is among the basal archosaur thecodonts. Some of the remains destroyed in World War II by Axis bombing. May be example of island dwarfism.

Nambalia roychowdhurii

—— 2.5 m (8 ft) TL, 20 kg (40 lb)
FOSSIL REMAINS Minority of two skeletons.
ANATOMICAL CHARACTERISTICS Insufficient information.
AGE Late Triassic, late Norian or earliest Rhaetian.
DISTRIBUTION AND FORMATION/S Central India; upper Maleri.

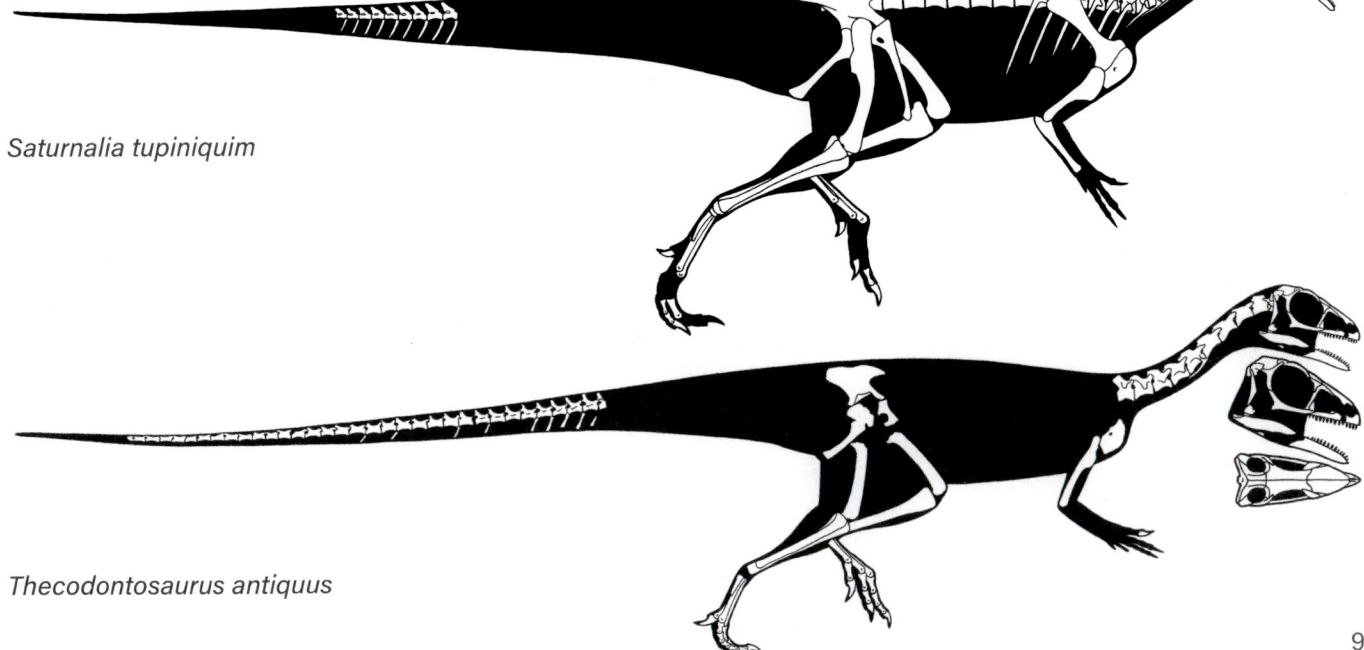

Saturnalia tupiniquim

Thecodontosaurus antiquus

EUPROSAUROPODS

SMALL TO LARGE HERBIVOROUS AND OMNIVOROUS PROSAUROPODS LIMITED TO THE LATE TRIASSIC AND EARLY JURASSIC, ALL CONTINENTS.

ANATOMICAL CHARACTERISTICS Stoutly built. Partial cheeks very probably present in at least some species, same for incipient beaks, antorbital fenestra reduced, teeth blunt. Necks moderately long, quite slender. Trunks long. Thumbs and claws enlarged. Large claw on innermost toes. Air-sac systems may be beginning to appear in some examples.
HABITS First herbivores able to high browse, especially when rearing.
NOTES Euprosauropods are prosauropods excluding sauropods with elongated, slender necks three-quarters or more the length of the presacrals that include *Plateosaurus*.

Macrocollum itaquii

—— 3.1 m (10 ft) TL, 90 kg (200 lb)
FOSSIL REMAINS Two nearly complete skulls and skeletons, other remains.
ANATOMICAL CHARACTERISTICS Head quite small, subrectangular in side view, broad in top view, including snout, snout a little downturned. Arm moderately long. Bi/quadrupedal.
AGE Late Triassic, early Norian.
DISTRIBUTION AND FORMATION/S Southern Brazil; Caturrita.

Unaysaurus tolentinoi

—— Adult size uncertain
FOSSIL REMAINS Nearly complete skull and minority of skeleton, minority of juvenile skeleton.

ANATOMICAL CHARACTERISTICS Head shallow, subrectangular, snout a little downturned. Arm moderately long. Bi/quadrupedal.
AGE Late Triassic, early Norian.
DISTRIBUTION AND FORMATION/S Southern Brazil; Caturrita.

Jaklapallisaurus asymmetricus

—— 3.5 m (11 ft) TL, 100 kg (200 lb)
FOSSIL REMAINS Minority of one skeleton.
ANATOMICAL CHARACTERISTICS Insufficient information.
AGE Late Triassic, late Norian or earliest Rhaetian.
DISTRIBUTION AND FORMATION/S Central India; upper Maleri.

Unaysaurus tolentinoi

Macrocollum itaquii

*Plateosaurus
trossingensis*

Plateosaurus (or *Issi*) *saaneq* (or *gracilis*)

—— Adult size uncertain
FOSSIL REMAINS Majority of two skulls, immature.
ANATOMICAL CHARACTERISTICS Similar to *P. gracilis*.
AGE Late Triassic, middle Norian.
DISTRIBUTION AND FORMATION/S Eastern Greenland; upper Malmros Klint.
NOTES May be *Plateosaurus*, specifically *P. gracilis*.

Plateosaurus (or *Sellosaurus*) *gracilis*

—— 5 m (15 ft) TL, 275 kg (600 lb)
FOSSIL REMAINS Majority of two dozen partial skulls and skeletons.
ANATOMICAL CHARACTERISTICS Head shallow, subrectangular, snout a little downturned. Arm moderately long. Bi/quadrupedal.
AGE Late Triassic, middle Norian.
DISTRIBUTION AND FORMATION/S Southern Germany; lower and middle Löwenstein.
NOTES *Efraasia diagnosticus* probably an immature form of this species. Maximum size uncertain. May be direct ancestor of *P. trossingensis*.

Plateosaurus trossingensis

—— 8.5 m (27 ft) TL, 2 tonnes
FOSSIL REMAINS Dozens of complete to partial skulls and skeletons, adult to juvenile, completely known.
ANATOMICAL CHARACTERISTICS Head shallow, narrow, elongated, especially snout, which is a little deeper than back of head, narrow, a little downturned. Skeleton heavily built. Arm moderately long. Bi/quadrupedal.
AGE Late Triassic, late Norian.

DISTRIBUTION AND FORMATION/S Germany, Switzerland, eastern France; Trossingen, Klettgau, Marnes Irisées Supérieures.
ONTOGENY Some individuals fully mature at about half size of other individuals, did not grow through life.
NOTES The classic prosauropod known from abundant remains. New large skulls demonstrate long-snouted shape of adult skulls. Probably includes *P. longiceps*, *P. engelhardti*, *P. erienbergensis*, *P. integer*, *P. frassianus*, *Ruehleia bedheimensis*, and *Tuebingosaurus maierfritzorum*. If the latter mature fossil is in this taxon, then indicates *Plateosaurus* is a more-derived prosauropod than commonly thought, may include similarly large *Gresslyosaurus ingens* etc. If larger than *P. gracilis*, may be because of bigger predators.

Plateosauravus cullingworthi

—— 9 m (30 ft) TL, 2.4 tonnes
FOSSIL REMAINS A few partial skeletons.
ANATOMICAL CHARACTERISTICS Insufficient information.
AGE Late Triassic, Rhaetian.
DISTRIBUTION AND FORMATION/S South Africa; lower Elliot.
HABITAT Arid.
NOTES Was *Euskelosaurus browni*, which is based on inadequate remains.

Kholumolumo ellenbergerorum

—— 7 m (23 ft) TL, 1.1 tonne
FOSSIL REMAINS Numerous disarticulated individuals from a bone bed.

Plateosaurus (or *Sellosaurus*) *gracilis*

immature

adult

Plateosaurus trossingensis

ANATOMICAL CHARACTERISTICS Insufficient
information.
AGE Late Triassic, Rhaetian.
DISTRIBUTION AND FORMATION/S South Africa; lower
Elliot.
HABITAT Arid.

Musankwa sanyatiensis
—— 4.5 m (15 ft) TL, 250 kg (500 lb)
FOSSIL REMAINS Minority of skeleton.
ANATOMICAL CHARACTERISTICS Insufficient
information.
AGE Late Triassic, Norian?

DISTRIBUTION AND FORMATION/S Zimbabwe; Pebbly
Arkose.

Massospondylus carinatus
—— 4.3 m (14 ft) TL, 195 kg (430 lb)
FOSSIL REMAINS Many dozens of skulls and skeletons,
many complete, juveniles to adult, completely known,
nests with up to three dozen eggs, embryos.
ANATOMICAL CHARACTERISTICS Head subrectangular,
fairly deep. Thumb and foot claws large. Arm long in
juveniles, moderately long in adults, indicating increasing
bipedalism with growth. Eggs spherical, 60 mm (2.4 in) in
diameter.

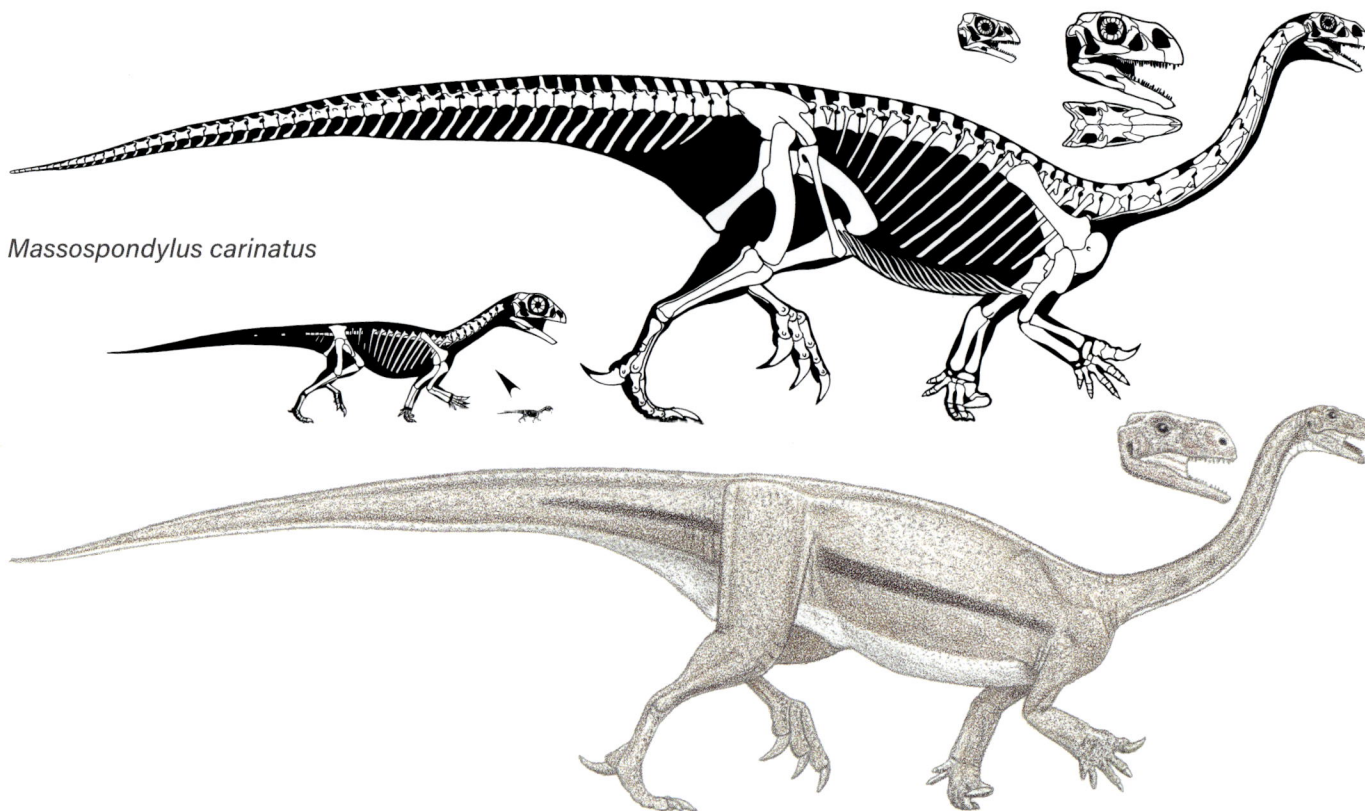

Massospondylus carinatus

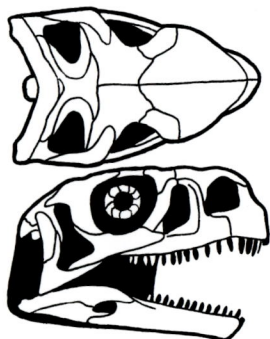

Ngwevu intloko

AGE Early Jurassic, late Hettangian to perhaps early Pliensbachian.
DISTRIBUTION AND FORMATION/S South Africa, Lesotho, Zimbabwe; upper Elliot, Bushveld Sandstone, Upper Karoo Sandstone, Forest Sandstone.
HABITAT In at least some locations, desert.
HABITS Probably fed on vegetation along watercourses and at oases. Eggs probably buried and abandoned.
NOTES The original fossil is inadequate, and the time span is suspiciously long for a single species. *Massospondylus kaalae* from the upper Elliot Formation may be a distinct species. *Ignavusaurus rachelis* may be a juvenile of this or another taxon from the upper Elliot.

Ngwevu intloko
—— 2 m (6 ft) TL, 20 kg (40 lb)
FOSSIL REMAINS Complete skull and partial skeleton.

ANATOMICAL CHARACTERISTICS Head short and deep mainly because snout is short and deep, subrectangular, broad.
AGE Early Jurassic, Sinemurian?
DISTRIBUTION AND FORMATION/S South Africa; uppermost Elliot.
HABITAT Arid.
HABITS Short, broad head suggests more powerful bite than usual in prosauropods for feeding on tougher vegetation.
NOTES Adult size apparently small, had been thought to be a juvenile *Massospondylus*.

Adeopapposaurus mognai
—— Adult size uncertain
FOSSIL REMAINS Majority of a few skulls and skeletons.
ANATOMICAL CHARACTERISTICS Head shallow, subrectangular, broad. Arm short. Strongly bipedal.
AGE Early Jurassic.
DISTRIBUTION AND FORMATION/S Southern Argentina; Cañón del Colorado.
NOTES Skeletal proportions not entirely certain.

Lufengosaurus huenei
—— 9 m (30 ft) TL, 1.7 tonnes
FOSSIL REMAINS Over two dozen skulls and skeletons, some complete, juvenile to adult, completely known.
ANATOMICAL CHARACTERISTICS Neck longer than that of most prosauropods. Arm short. Strongly bipedal.

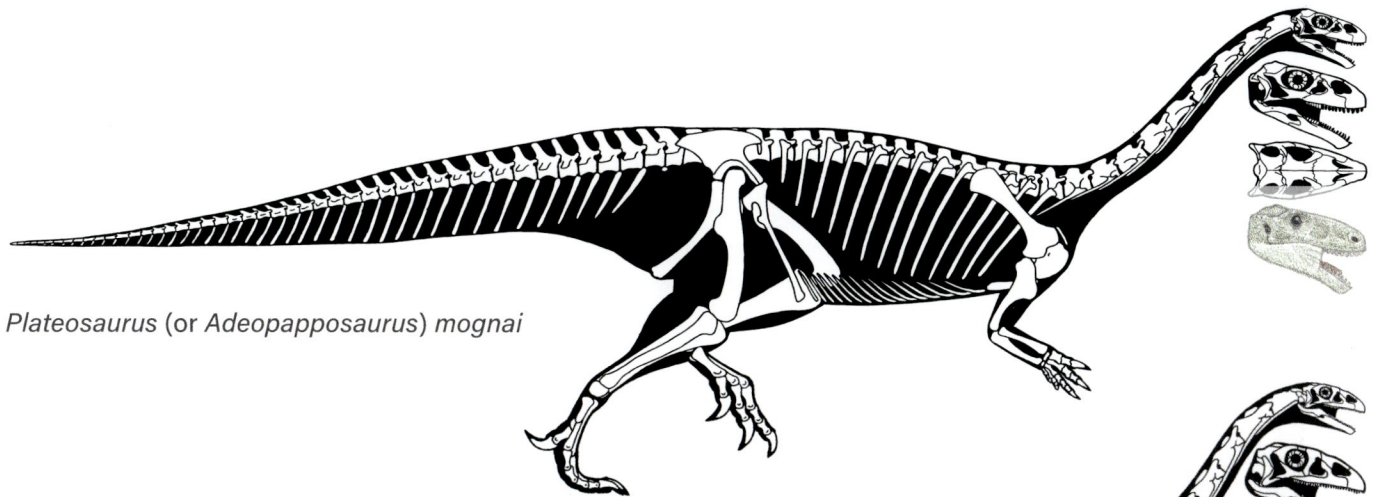

Plateosaurus (or *Adeopapposaurus*) *mognai*

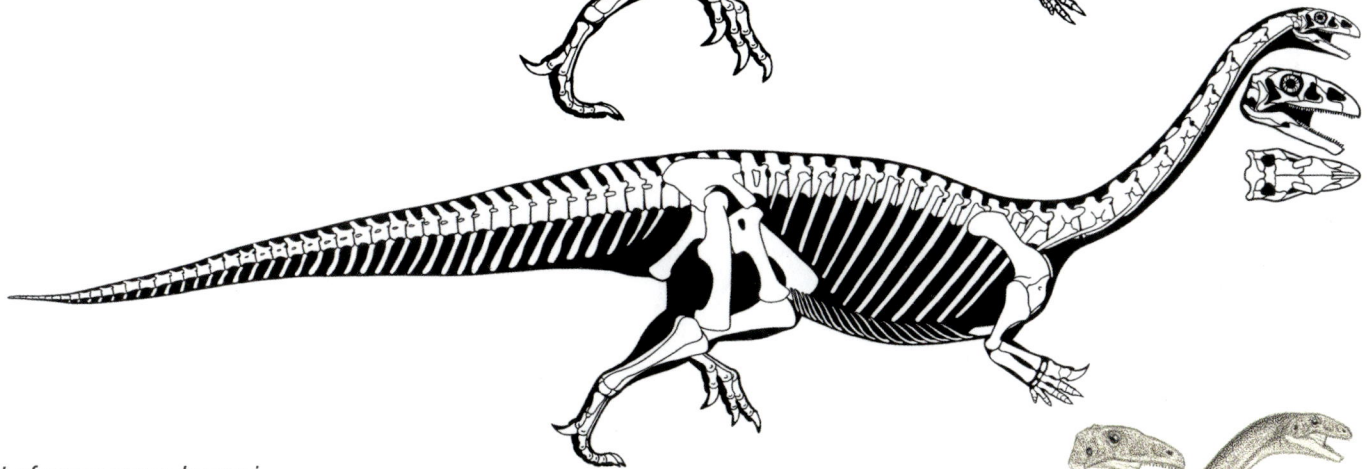

Lufengosaurus huenei

AGE Early Jurassic, Sinemurian.
DISTRIBUTION AND FORMATION/S Southwestern China; upper Lufeng.
NOTES *L. magnus* probably adult of this species, *Gyposaurus sinensis* probably juveniles.

Leyesaurus marayensis
—— 3 m (10 ft) TL, 65 kg (140 lb)
FOSSIL REMAINS Majority of skull and minority of skeleton.

Leyesaurus marayensis

ANATOMICAL CHARACTERISTICS Head shallow, subrectangular, broad.
AGE Early Jurassic?, Hettangian?
DISTRIBUTION AND FORMATION/S Northwestern Argentina; uppermost Quebrada del Barro.
HABITAT Semiarid.

Sarahsaurus aurifontanalis
—— 4 m (13 ft) TL, 180 kg (400 lb)
FOSSIL REMAINS Majority of skull and two skeletons.
ANATOMICAL CHARACTERISTICS Head shallow, snout probably a little downturned. Arm short. Strongly bipedal.
AGE Early Jurassic.
DISTRIBUTION AND FORMATION/S Arizona; Kayenta.
HABITAT Semiarid.

Lufengosaurus huenei

Sarahsaurus aurifontanalis

Coloradisaurus brevis

—— 3 m (10 ft) TL, 65 kg (140 lb)
FOSSIL REMAINS Complete skull.
ANATOMICAL CHARACTERISTICS Head short,
subtriangular, broad.
AGE Late Triassic, middle Norian.
DISTRIBUTION AND FORMATION/S Northern Argentina;
Los Colorados.
HABITAT Seasonally wet woodlands.

Coloradisaurus brevis

Xingxiulong chengi

—— 6 m (20 ft) TL, 500 kg (1,000 lb)
FOSSIL REMAINS Two partial skulls and majority of
skeletons of adult, majority of juvenile skeleton.
ANATOMICAL CHARACTERISTICS Robustly built.
Strongly bipedal.
AGE Early Jurassic, Hettangian.

DISTRIBUTION AND FORMATION/S Southwestern China;
lower Lufeng.
NOTES May be direct ancestor of *X. yueorum*.

Xingxiulong? yueorum

—— 8 m (25 ft) TL, 1.4 tonnes
FOSSIL REMAINS Majority of skeleton.
ANATOMICAL CHARACTERISTICS Robustly built.
Strongly bipedal.
AGE Early Jurassic, Sinemurian.
DISTRIBUTION AND FORMATION/S Southwestern China;
upper Lufeng.

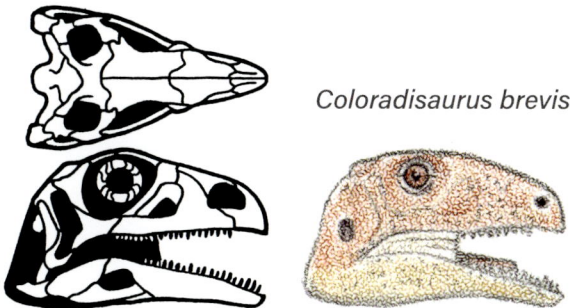

Jingshanosaurus xinwaensis

—— 9 m (30 ft) TL, 1.7 tonnes
FOSSIL REMAINS Complete skull and skeleton.
ANATOMICAL CHARACTERISTICS Head subtriangular,
broad, cheeks may have been absent. Arm short.
Strongly bipedal.
AGE Early Jurassic, late Hettangian and/or early
Sinemurian.
DISTRIBUTION AND FORMATION/S Southwestern China;
upper lower Lufeng.
NOTES *Chuxiongosaurus lufengensis* and *Xixiposaurus suni*
may be juveniles of this taxon.

Jingshanosaurus xinwaensis
(see also next page)

Jingshanosaurus xinwaensis

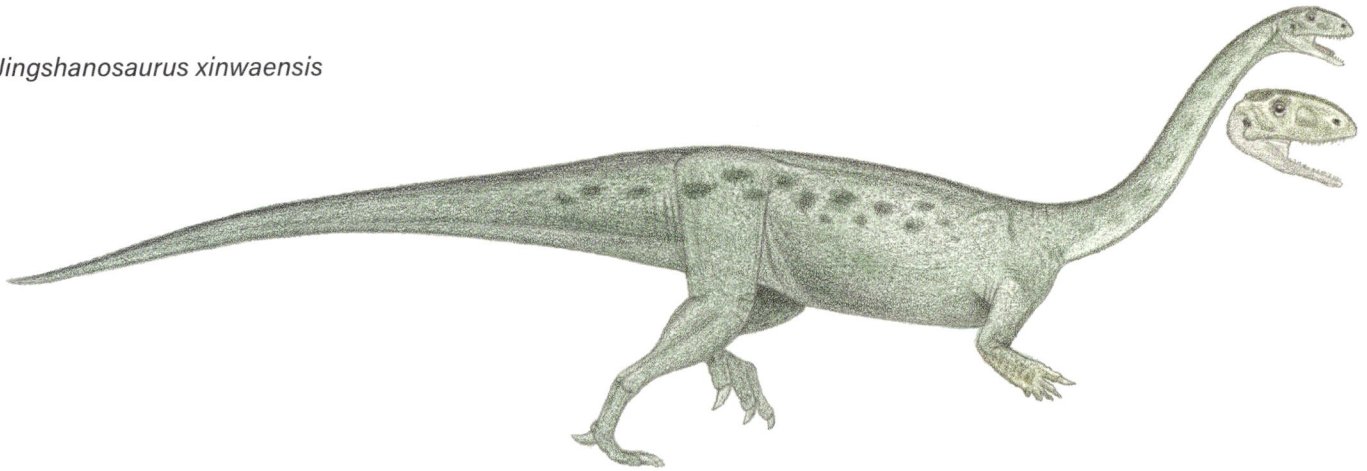

Lishulong wangi
—— 8 m (26 ft) TL, 1.4 tonnes
FOSSIL REMAINS Majority of skull and minority of skeleton.
ANATOMICAL CHARACTERISTICS Head apparently deep.
AGE Early Jurassic, late Hettangian and/or early Sinemurian.
DISTRIBUTION AND FORMATION/S Southwestern China; upper lower Lufeng.

Yunnanosaurus huangi
—— 4.5 m (15 ft) TL, 240 kg (500 lb)
FOSSIL REMAINS Almost two dozen skulls and skeletons, some complete, juvenile to adult.
ANATOMICAL CHARACTERISTICS Head small, subtriangular, cheeks may have been absent. Arm short. Strongly bipedal.
AGE Early Jurassic, Sinemurian.
DISTRIBUTION AND FORMATION/S Southwestern China; upper Lufeng.
NOTES *Y. robustus* is probably the adult of this species.

Qianlong shouhu
—— 7 m (23 ft) TL, 1,000 kg (2,000 lb)
FOSSIL REMAINS Three partial adults, egg clutches, some with embryos.
ANATOMICAL CHARACTERISTICS Fairly robust. Snout short and subtriangular. Adults strongly bipedal.
AGE Early Jurassic, early Sinemurian.
DISTRIBUTION AND FORMATION/S Southern China; lower Ziliujing.

Yimenosaurus youngi
—— 9 m (30 ft) TL, 2 tonnes
FOSSIL REMAINS Majority of skull, numerous partial skulls and skeletons.
ANATOMICAL CHARACTERISTICS Head deep, subrectangular.

Yimenosaurus youngi

Yunnanosaurus huangi

AGE Early Jurassic, Pliensbachian.
DISTRIBUTION AND FORMATION/S Southwestern China; upper? Fengjiahe.
NOTES Was thought to be from Middle Jurassic. *Irisosaurus yimenensis* may be an immature example of this species.

Seitaad ruessi

—— 3 m (10 ft) TL, 65 kg (140 lb)
FOSSIL REMAINS Minority of skeleton.
ANATOMICAL CHARACTERISTICS Arm moderately short. Predominantly bipedal.
AGE Early Jurassic, Pliensbachian.
DISTRIBUTION AND FORMATION/S Utah; lower Navajo Sandstone.
HABITAT Dune desert.

Eucnemesaurus fortis

—— 8 m (26 ft) TL, 1.4 tonnes
FOSSIL REMAINS Small portion of a few skeletons.
ANATOMICAL CHARACTERISTICS Insufficient information.
AGE Late Triassic, Rhaetian.
DISTRIBUTION AND FORMATION/S South Africa; lower Elliot.
HABITAT Arid.
NOTES The scanty remains were once labeled *Aliwalia rex*, which was thought to be a giant baso-theropod. *Eucnemesaurus entaxonis* is probably the adult of this species.

Mussaurus patagonicus

—— 6.5 m (20 ft) TL, 500 kg (1,100 lb)
FOSSIL REMAINS Many dozens of skulls and skeletons from majority to partial, mainly juveniles, including embryos and hatchlings up to adults, many dozens of eggs.
ANATOMICAL CHARACTERISTICS Quadrupedal to more bipedal with growth.
AGE Late Triassic, Norian.
DISTRIBUTION AND FORMATION/S Southern Argentina; Laguna Colorada.
HABITS Small juveniles may have supplemented diet with insects.
NOTES Partial nature of most remains and incomplete description hinder restorations.

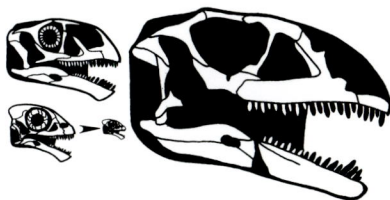

Mussaurus patagonicus skulls growth series

Sefapanosaurus zastronensis

—— 4 m (9 ft) TL, 200 kg (400 lb)
FOSSIL REMAINS Four partial skeletons.
ANATOMICAL CHARACTERISTICS Insufficient information.
AGE Late Triassic, Norian.
DISTRIBUTION AND FORMATION/S South Africa; Elliot, level uncertain.
HABITAT Arid.

Camelotia borealis

—— 10 m (33 ft) TL, 2.5 tonnes
FOSSIL REMAINS Minority of skeleton.
ANATOMICAL CHARACTERISTICS Insufficient information.
AGE Late Triassic, Rhaetian.
DISTRIBUTION AND FORMATION/S Southwestern England; Westbury.

Aardonyx celestae

—— Adult size uncertain
FOSSIL REMAINS Two or three partial skulls and skeletons.
ANATOMICAL CHARACTERISTICS Insufficient information.
AGE Early Jurassic, Sinemurian or Pliensbachian.
DISTRIBUTION AND FORMATION/S South Africa; upper Elliot.
HABITAT Arid.
NOTES *Arcusaurus pereirabdalorum* may be a juvenile of this species.

Leonerasaurus taquetrensis

—— 3 m (10 ft) TL, 70 kg (150 lb)
FOSSIL REMAINS Minority of skull and skeleton.
ANATOMICAL CHARACTERISTICS Insufficient information.
AGE Early Jurassic, late Sinemurian or Pliensbachian.
DISTRIBUTION AND FORMATION/S Southern Argentina; Las Leoneras.

Riojasaurus incertus

—— 6.6 m (22 ft) TL, 860 kg (1,900 lb)
FOSSIL REMAINS Complete skull, numerous skeletons of varying completeness, juvenile to adult.
ANATOMICAL CHARACTERISTICS Head subtriangular. Arm long, robust. Strongly quadrupedal.

Riojasaurus incertus

AGE Late Triassic, middle Norian.
DISTRIBUTION AND FORMATION/S Northern Argentina; Los Colorados.
HABITAT Seasonally wet woodlands.

Melanorosaurus readi
—— 6 m (20 ft) TL, 500 kg (1,000 lb)
FOSSIL REMAINS Minority of skeleton.
ANATOMICAL CHARACTERISTICS Insufficient information.
AGE Late Triassic, Rhaetian.
DISTRIBUTION AND FORMATION/S South Africa, lower Elliot.
HABITAT Arid.
NOTES Below specimens probably do not belong to this taxon.

Unnamed genus and species
—— 8 m (26 ft) TL, 1.5 tonnes
FOSSIL REMAINS Complete distorted skull and majority of skeleton, partial remains.

ANATOMICAL CHARACTERISTICS Head strongly constructed. Arm long, robust. Strongly quadrupedal.
AGE Late Triassic, Rhaetian.
DISTRIBUTION AND FORMATION/S South Africa, lower Elliot.
HABITAT Arid.
NOTES Usual placement in *Melanorosaurus* under question.

Meroktenos thabanensis
—— Adult size uncertain
FOSSIL REMAINS Minority of skeleton.
ANATOMICAL CHARACTERISTICS Robustly built.
AGE Late Triassic, Rhaetian.
DISTRIBUTION AND FORMATION/S South Africa, lower Elliot.
HABITAT Arid.
NOTES Was *Melanorosaurus thabanensis*.

Unnamed genus and species

Unnamed genus and species

Unnamed genus *youngi*
—— 10.5 m (40 ft) TL, 3.5 tonnes
FOSSIL REMAINS Partial skeleton.
ANATOMICAL CHARACTERISTICS Neck moderately long. Tail base deep.
AGE Early Jurassic, Pliensbachian.
DISTRIBUTION AND FORMATION/S Southwestern China; Fengjiahe.
NOTES Placement in earlier *Yunnanosaurus* highly problematic. Largest known prosauropod, but not by as much as some past estimates.

Anchisaurus polyzelus
—— 2.1 m (7 ft) TL, 21 kg (45 lb)
FOSSIL REMAINS Nearly complete skull and majority of skeleton.
ANATOMICAL CHARACTERISTICS Head shallow, subtriangular, broad. Arm moderately long. Bi/quadrupedal.
AGE Early Jurassic, Hettangian or Sinemurian.
DISTRIBUTION AND FORMATION/S Connecticut, Massachusetts; Portland.
HABITAT Semiarid rift valley with lakes.
NOTES *Ammosaurus major* is probably the adult of this species.

Blikanasaurus cromptoni
—— 4 m (13 ft) TL, 200 kg (450 lb)
FOSSIL REMAINS Minority of skeleton.
ANATOMICAL CHARACTERISTICS Leg massively built.
AGE Late Triassic, Rhaetian.
DISTRIBUTION AND FORMATION/S South Africa; lower Elliot.
HABITAT Arid.

Yizhousaurus sunae
—— 7 m (23 ft) TL, 1.2 tonnes
FOSSIL REMAINS Majority of skull and skeleton.
ANATOMICAL CHARACTERISTICS Head short, deep, broad. Strongly bipedal.
AGE Early Jurassic, Sinemurian.
DISTRIBUTION AND FORMATION/S Southwestern China; upper Lufeng.
HABITS Short, broad head suggests more powerful bite than usual in prosauropods for feeding on tougher vegetation.
NOTES Combines a somewhat sauropod-type skull with prosauropod-grade skeleton. Aft neck and front truck vertebra are so wedge-shaped that it is difficult to position them normally, in life, thick, counter-wedge-shaped cartilage pads filled out the spaces.

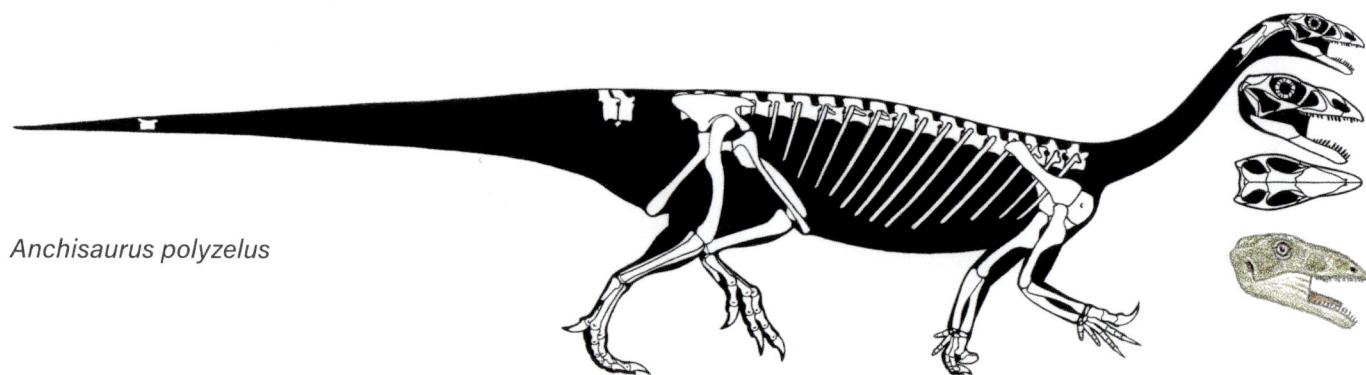

Anchisaurus polyzelus

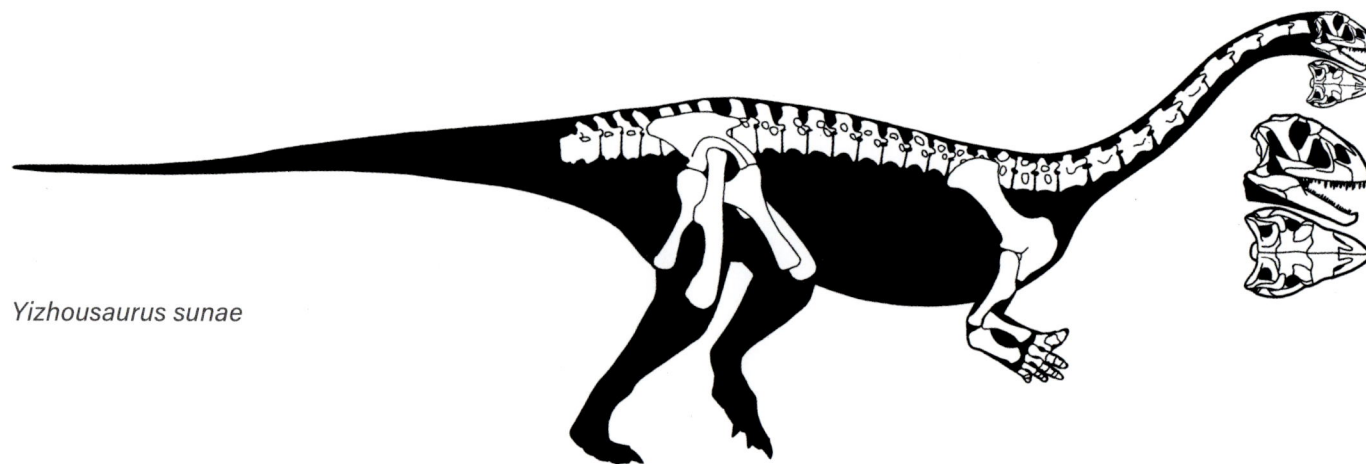

Yizhousaurus sunae

Yizhousaurus sunae

Unnamed genus and species
—— 4.5 m (18 ft) TL, 250 kg (500 lb)
FOSSIL REMAINS Minority of skeleton.
ANATOMICAL CHARACTERISTICS Insufficient
information.
AGE Early Jurassic, Pliensbachian or Toarcian.
DISTRIBUTION AND FORMATION/S Arizona; Navajo
Sandstone.

HABITAT Dune desert.
HABITS Probably fed on vegetation along watercourses
and at oases.
NOTES Has been placed in earlier and geographically
distant *Ammosaurus* and *Massospondylus*, both options
highly problematic.

SAUROPODOMORPH MISCELLANEA
NOTES These may be the most-derived known prosauropods, or the basalmost known sauropods.

Lessemsaurus sauropoides
—— 11 m (35 ft) TL, 3 tonnes
FOSSIL REMAINS Minority of four skeletons.
ANATOMICAL CHARACTERISTICS Insufficient
information.
AGE Late Triassic, middle Norian.
DISTRIBUTION AND FORMATION/S Northern Argentina;
upper Los Colorados.
HABITAT Seasonally wet woodlands.
NOTES Markedly higher mass estimates are not correct.

Ingentia prima
—— 10 m (33 ft) TL, 2.5 tonnes
FOSSIL REMAINS Minority of two skeletons.
ANATOMICAL CHARACTERISTICS Neck moderately long.
AGE Late Triassic, late Norian or Rhaetian.

DISTRIBUTION AND FORMATION/S Northwestern
Argentina; Quebrada del Barro.
NOTES Markedly higher mass estimates are not correct.

Antetonitrus ingenipes
—— 13 m (43 ft) TL, 7 tonnes
FOSSIL REMAINS Minority of skeletons.
ANATOMICAL CHARACTERISTICS Insufficient
information.
AGE Early Jurassic, Hettangian or Sinemurian.
DISTRIBUTION AND FORMATION/S South Africa; upper
Elliot.
HABITAT Arid.
NOTES *Ledumahadi mafube* may be adult of this species.
If so, is earliest known giant dinosaur. Higher mass
estimates are problematic.

SAUROPODS

LARGE TO ENORMOUS HERBIVOROUS SAUROPODOMORPHS FROM THE EARLY JURASSIC TO THE END OF THE DINOSAUR ERA, MOST CONTINENTS.

ANATOMICAL CHARACTERISTICS Variable. Skulls and teeth more heavily built than those of prosauropods, nostrils at least somewhat retracted. Skeletons heavily built. Neck moderately to extremely long, with over 10 cervicals. Tails moderately to extremely long. Skeletons at least incipiently pneumatic, birdlike, some degree of air sac–ventilated respiratory system present. Quadrupedal when moving normally, arm and leg less flexed than in prosauropods. Lower legs shorter than upper, feet short and broad. Could not achieve full run with suspended phase with all feet off ground at same time.
HABITATS Very variable, deserts to well-watered forests, tropics to subpolar regions, apparent preference for seasonally dry/wet woodlands.
HABITS Highest browsers to have evolved. Preferred tough vegetation, especially conifers. Hard-pressed or unable to completely submerge because of pneumatic bodies.
NOTES Lasting 140 million years or more and regularly rivaling great whales in size, the most successful group of large herbivorous animals that has yet evolved. Trackways from Greenland suggest first examples evolved in Early Jurassic. Absence from Antarctica probably reflects lack of sufficient sampling. Never small, known adults always at least close to 2 tonnes, even when island dwellers, probably partly because columnar limbs and resulting slow top speeds forced large size to deal with predators.

BASO-SAUROPODS

LARGE TO GIGANTIC SAUROPODS LIMITED TO THE EARLY JURASSIC OF THE EASTERN HEMISPHERE.

ANATOMICAL CHARACTERISTICS Fairly uniform. Heads short, snouts narrow and rounded. Necks and tails moderately long. Limbs moderately flexed. Arm moderately long, so shoulders about as high as hips. Hands not forming an arcade, fingers not extremely abbreviated. Pelvic ilia shallow, ankles still markedly flexible. Skeletal pneumaticity partly developed, so birdlike respiratory system developing.
HABITS Probably feeding generalists. Probably able to run slowly. Main defense standing and fighting with claws.
ENERGETICS Thermophysiology probably intermediate between that of prosauropods and eusauropods.
NOTES This group is a grade splittable into more divisions. Various anatomical features remain poorly understood. If Late Triassic trackways are those of basal-most sauropods, then small-sized examples had appeared by then. Absence from the Western Hemisphere may reflect lack of sufficient sampling.

Chinshakiangosaurus chunghoensis
—— 10 m (30 ft) TL, 3 tonnes
FOSSIL REMAINS Minority of skull and skeleton.
ANATOMICAL CHARACTERISTICS Mouth fairly broad, extensive cheeks present.
AGE Early Jurassic.
DISTRIBUTION AND FORMATION/S Southern China; Fengjiahe.

Ohmdenosaurus liasicus
—— Adult size uncertain
FOSSIL REMAINS Minority of skeleton, possibly juvenile.
ANATOMICAL CHARACTERISTICS Insufficient information.
AGE Early Jurassic, early Toarcian.
DISTRIBUTION AND FORMATIONS Southern Germany; lower Posidonienschiefer.
HABITAT Island archipelago shallows.
HABITS Hunter of small and medium-sized game.
NOTES Found as drift in interisland deposits.

Gongxianosaurus shibeiensis
—— 11 m (35 ft) TL, 3.2 tonnes
FOSSIL REMAINS Majority of skeleton.
ANATOMICAL CHARACTERISTICS Arm less prosauropod-like. Base of tail deep.
AGE Early Jurassic, Toarcian.
DISTRIBUTION AND FORMATION/S Central China; Ziliujing.

Vulcanodon karibaensis
—— 11 m (35 ft) TL, 3.5 tonnes
FOSSIL REMAINS Minority of skeleton.
ANATOMICAL CHARACTERISTICS Standard for group.
AGE Early Jurassic, late Pliensbachian or Toarcian.
DISTRIBUTION AND FORMATION/S Zimbabwe; Batoka.
HABITAT Arid.

Pulanesaura eocollum
—— 11 m (35 ft) TL, 3.5 tonnes
FOSSIL REMAINS Minority of skeleton.

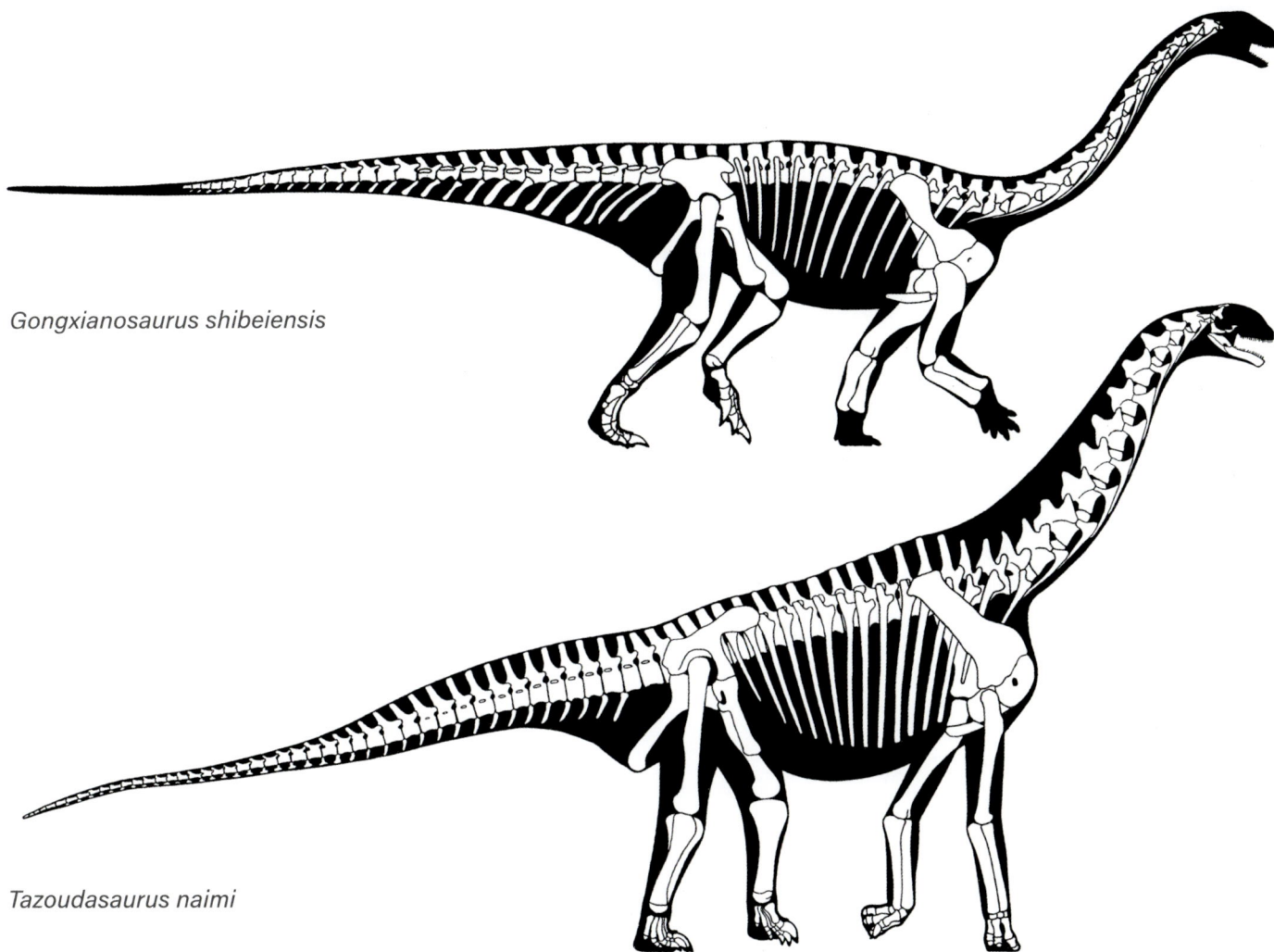

Gongxianosaurus shibeiensis

Tazoudasaurus naimi

ANATOMICAL CHARACTERISTICS Standard for group.
AGE Early Jurassic, late Hettangian or Sinemurian.
DISTRIBUTION AND FORMATION/S South Africa; upper Elliot.
HABITAT Arid.

Tazoudasaurus naimi
—— 10 m (33 ft) TL, 3.7 tonnes
FOSSIL REMAINS Minority of skull and two skeletons, juvenile and adult.
ANATOMICAL CHARACTERISTICS Neck not elongated. Shoulder somewhat higher than hip.
AGE Early Jurassic, Toarcian.
DISTRIBUTION AND FORMATION/S Morocco; Douar of Tazouda.

Kotasaurus yamanpalliensis
—— 9 m (30 ft) TL, 2.5 tonnes
FOSSIL REMAINS Majority of skeleton.
ANATOMICAL CHARACTERISTICS Tail tipped by small, roundish club.
AGE Early Jurassic.

DISTRIBUTION AND FORMATION/S Southeastern India; Kota.
HABITS Defense included high-velocity impacts from tail club, which was probably also used in intraspecific combat.

Isanosaurus attavipachi
—— 13 m (43 ft) TL, 7 tonnes
FOSSIL REMAINS Minority of several skeletons, juvenile and adult.
ANATOMICAL CHARACTERISTICS Insufficient information.
AGE Jurassic.
DISTRIBUTION AND FORMATION/S Thailand; Nam Phong.
NOTES The large remains are probably the adult of the juvenile remains named *Isanosaurus*. Originally incorrectly thought to be from the Late Triassic and thereby showing evolution of giant sauropods soon after first appearance of dinosaurs, instead may be from Late Jurassic.

EUSAUROPODS

LARGE TO ENORMOUS SAUROPODS FROM THE EARLY JURASSIC TO THE END OF THE DINOSAUR ERA, MOST CONTINENTS.

ANATOMICAL CHARACTERISTICS Fairly variable. Snouts broader, rounded or squared off, nostrils further retracted, cheeks absent. Skeletons massively built. Necks moderately to extremely long. Trunks compact, deep, vertebral series usually stiffened. Tails moderately to extremely long. Shoulder girdles large, hand forming a vertical arcade. Fingers very short and rigid or lost, no padding, large claw limited to thumb or lost. Pelves large, ilia deep and strongly arced at top, indicating enlarged upper leg muscles, pubes subvertical. Lower legs shorter than upper, mobility of ankle limited. Feet very short and broad, five toes short and underlain by large pad, inner toes bearing large claws increasing in size progressing inward. Quadrupedal when moving normally, arms and legs columnar and massively built, so unable to amble faster than elephants. Skeletal pneumaticity and birdlike respiratory system better developed. Skin consisting of small scales often in rosette patterns.

ONTOGENY Growth rates moderate in at least some smaller species to moderately rapid, especially in giants, life spans may have approached and exceeded 100 years.

HABITATS Seasonally dry, open woodlands and prairies, and coastal wetlands, from tropics to polar regions.

HABITS High-level browsers and low-level grazers. Too slow to flee attackers, main defense standing and lashing out with clawed hands and feet or swinging tail, which often weighed tonnes and matched giant attacking theropods in mass. Long, tall necks may have been used for competitive display within species, delicate construction indicates they were not used as impact weapons for combat within species like giraffe necks. Trackways indicate that small juveniles formed pods of similarly sized individuals separate from the herds of large juveniles and adults over 1 tonne. Numerous trackways laid down along watercourses show that many sauropods of all sizes used shorelines to travel, but ability to move into water was limited because the narrow, padless hands were in danger of getting bogged down in soft sediments, as appears to have happened in some fossils. Probably used clawed hindfeet to dig for water in streambeds during droughts.

ENERGETICS Power production probably unusually high in longer-necked examples so that powerful heart could pump blood at very high pressures up to high-held brains.

NOTES The dinosaurs most similar to elephants and giraffes. Fragmentary remains and trackways indicate that some eusauropods approached and exceeded 100 tonnes, may have reached 200 tonnes.

BASO-EUSAUROPODS

LARGE EUSAUROPODS LIMITED TO THE MIDDLE TO LATE JURASSIC.

ANATOMICAL CHARACTERISTICS Heads short, snouts rounded. Necks rather short to moderately long, able to elevate subvertically. Tails moderately long. Arms moderately long, so shoulders as high as hips.

HABITS Low- to mid-level browsers.

NOTES Geographic distribution not yet known.

Shunosaurus lii (see p.110)
—— 9.5 m (30 ft) TL, 3.3 tonnes
FOSSIL REMAINS Numerous skulls and skeletons, completely known.
ANATOMICAL CHARACTERISTICS Neck short by sauropod standards. Sled-shaped chevrons under tail facilitated static rearing with tail as a prop. Tail tipped by small, spiked club. Arm and especially leg long, shoulders about as high as hips.
AGE Middle Jurassic, probably Bathonian.
DISTRIBUTION AND FORMATION/S Central China; lower Xiashaximiao.
HABITAT Heavily forested.
HABITS Fed at medium heights. Defense included high-velocity impacts from tail club, which was probably also used in intraspecific combat.
NOTES Almost as short-necked as *Brachytrachelopan*.

Unnamed genus? *jiangyiensis*
—— 15 m (50 ft) TL, 13 tonnes
FOSSIL REMAINS Partial skeleton.
ANATOMICAL CHARACTERISTICS Leg long.
AGE Middle Jurassic, probably Callovian.
DISTRIBUTION AND FORMATION/S Central China; upper Xiashaximiao.
HABITAT Heavily forested.
NOTES Placement in earlier, smaller *Shunosaurus* problematic.

Shunosaurus lii and *Gasosaurus constructus*

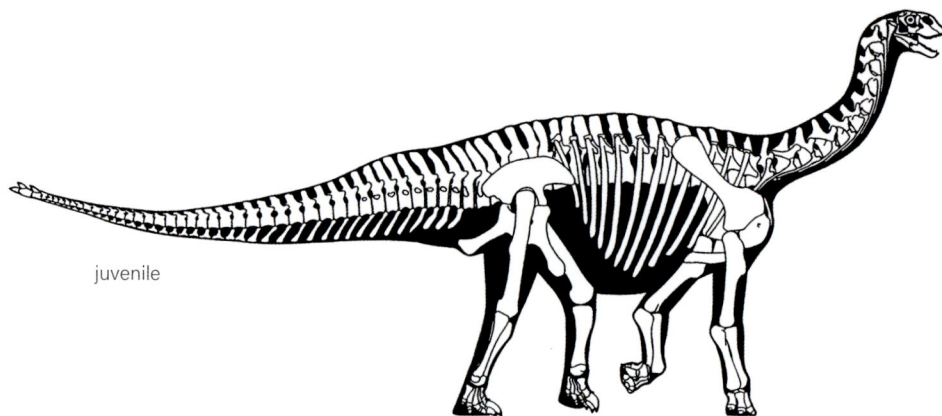

juvenile

adult

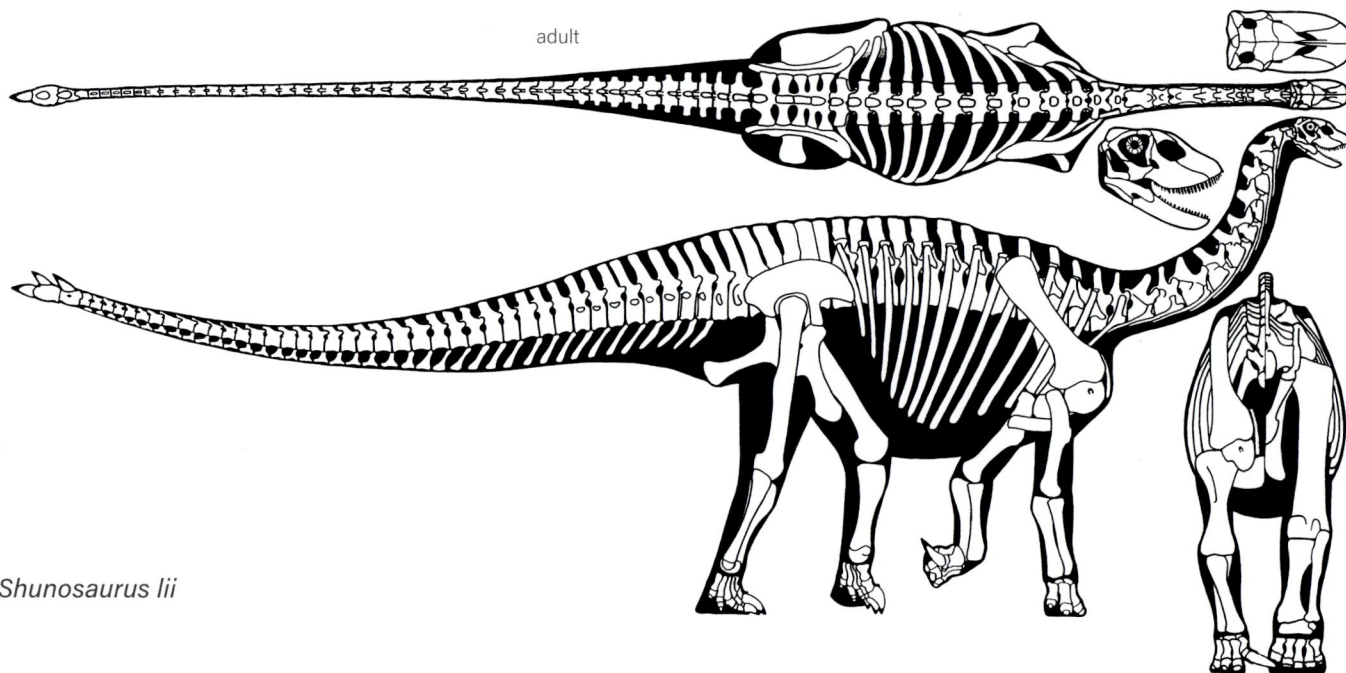

Shunosaurus lii

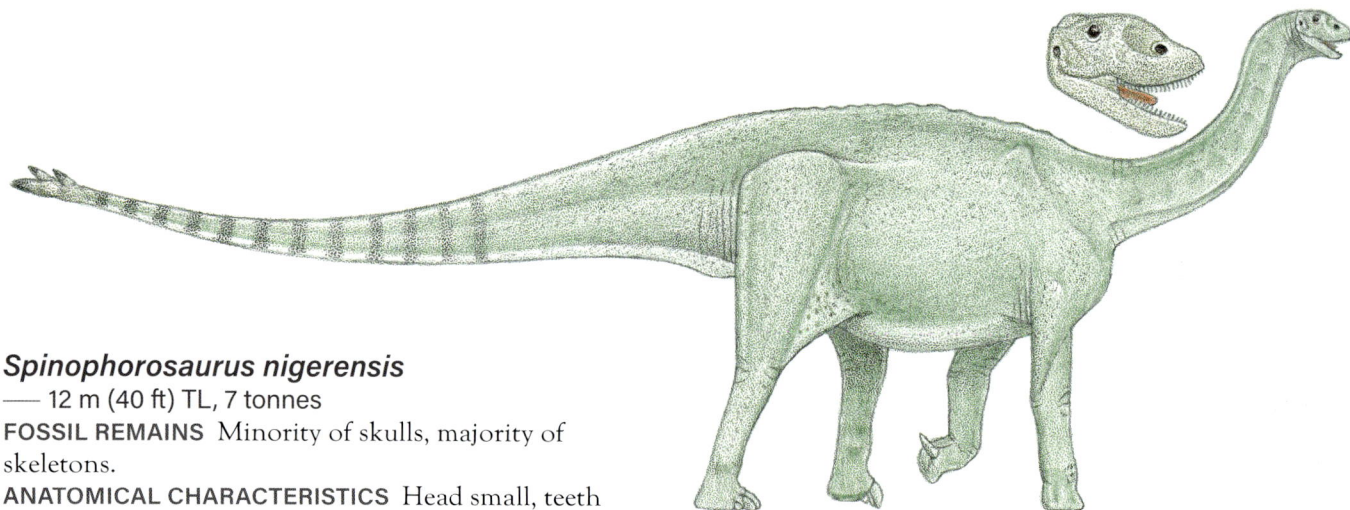

Spinophorosaurus nigerensis
—— 12 m (40 ft) TL, 7 tonnes
FOSSIL REMAINS Minority of skulls, majority of skeletons.
ANATOMICAL CHARACTERISTICS Head small, teeth stout. Neck moderately long. Tail base deep, sled-shaped chevrons under tail facilitated static rearing with tail as a prop. Shoulders appear to have been moderately high.
AGE Probably Middle Jurassic.
DISTRIBUTION AND FORMATION/S Niger; Irhazer.

HABITS Adapted for high browsing, both when quadrupedal and when bipedal or tripodal.
NOTES Restorations showing either level trunk or very high shoulders, and tail spikes incorrect.

Spinophorosaurus nigerensis

"CETIOSAURS"

LARGE TO GIGANTIC EUSAUROPODS LIMITED TO THE JURASSIC OF THE NORTHERN AND SOUTHERN HEMISPHERES.

ANATOMICAL CHARACTERISTICS Fairly uniform. Heads short, snouts rounded. Necks moderately long, able to elevate subvertically. Tails moderately long. Arms moderately long, so shoulders about as high as hips.
HABITS Probably feeding generalists.
NOTES Largely a grade of basal eusauropods, the relationships of many of these sauropods are uncertain.

Barapasaurus tagorei
—— 12 m (40 ft) TL, 7 tonnes
FOSSIL REMAINS Majority of skeleton from bone beds.
ANATOMICAL CHARACTERISTICS Neck moderately long.
AGE Early or middle Early Jurassic, Toarcian or Aalenian.
DISTRIBUTION AND FORMATION/S Southeastern India; lower Kota.

Dystrophaeus viaemalae
—— 13 m (43 ft) TL, 7 tonnes
FOSSIL REMAINS Minority of skeleton.
ANATOMICAL CHARACTERISTICS Insufficient information.
AGE Middle and/or Late Jurassic, Callovian and/or Oxfordian.
DISTRIBUTION AND FORMATION/S Utah; Summerville.
NOTES The relationships of *Dystrophaeus* are uncertain.

Barapasaurus tagorei

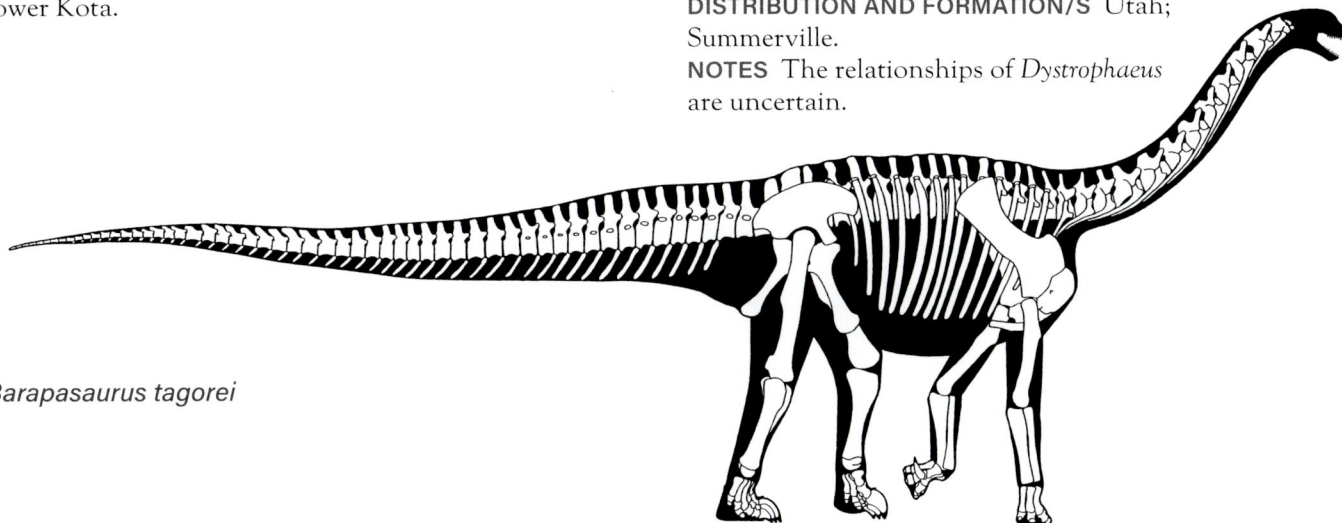

Rhoetosaurus brownei

—— 15 m (50 ft) TL, 9 tonnes
FOSSIL REMAINS Minority of skeleton.
ANATOMICAL CHARACTERISTICS Insufficient information.
AGE Middle Jurassic, Bajocian.
DISTRIBUTION AND FORMATION/S Northeastern Australia; Hutton.
HABITAT Polar forests with warm, daylight-dominated summers and cold, dark winters.

Datousaurus bashanensis

—— 10 m (34 ft) TL, 4.7 tonnes
FOSSIL REMAINS Partial skull and skeletons.
ANATOMICAL CHARACTERISTICS Neck moderately long. Shoulder a little higher than hip.
AGE Middle Jurassic, probably Bathonian.
DISTRIBUTION AND FORMATION/S Central China; lower Xiashaximiao.
HABITAT Heavily forested.
HABITS High-level browser.

Amygdalodon patagonicus

—— 12 m (40 ft) TL, 5 tonnes
FOSSIL REMAINS Minority of skeleton.
ANATOMICAL CHARACTERISTICS Insufficient information.
AGE Middle Jurassic, Bajocian.
DISTRIBUTION AND FORMATION/S Southern Argentina; Cerro Carnerero.

Bagualia alba

—— 12 m (40 ft) TL, 4 tonnes
FOSSIL REMAINS Partial skulls and skeletons.
ANATOMICAL CHARACTERISTICS Neck moderately long.
AGE Early Jurassic, early middle Toarcian.
DISTRIBUTION AND FORMATION/S Southern Argentina; lower Cañadón Asfalto.
HABITAT Short wet season, otherwise semiarid, riverine forests, open floodplains.

Patagosaurus fariasi

—— 16.5 m (53 ft) TL, 9.3 tonnes
FOSSIL REMAINS Minority of skull and numerous skeletons.
ANATOMICAL CHARACTERISTICS Neck moderately long. Tail long.
AGE Middle or Late Jurassic, late Callovian and/or early Oxfordian.
DISTRIBUTION AND FORMATION/S Southern Argentina; upper Cañadón Asfalto.
HABITAT Short wet season, otherwise semiarid, riverine forests, open floodplains.
HABITS Long tail facilitated rearing for high browsing.

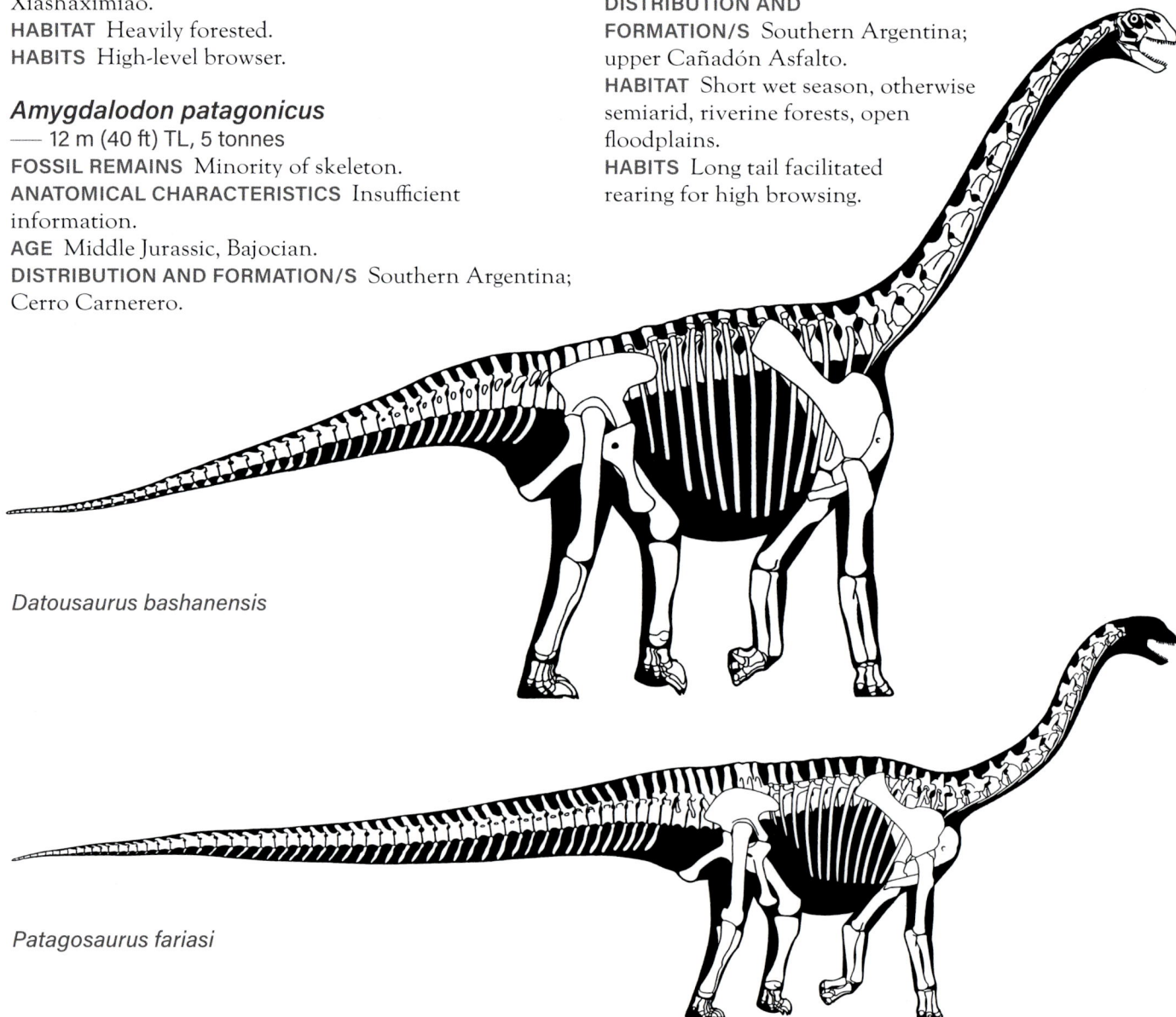

Datousaurus bashanensis

Patagosaurus fariasi

Volkheimeria chubutensis
—— Adult size uncertain
FOSSIL REMAINS Minority of skeleton, juvenile.
ANATOMICAL CHARACTERISTICS Insufficient information.
AGE Early Jurassic, middle Toarcian.
DISTRIBUTION AND FORMATION/S Southern Argentina; upper Cañadón Asfalto.

Chebsaurus algeriensis
—— Adult size uncertain
FOSSIL REMAINS Two partial skeletons, juvenile.
ANATOMICAL CHARACTERISTICS Insufficient information.
AGE Middle Jurassic, probably Callovian.
DISTRIBUTION AND FORMATION/S Algeria; unnamed.

Ferganasaurus verzilini
—— 18 m (60 ft) TL, 15 tonnes
FOSSIL REMAINS Minority of skeleton.
ANATOMICAL CHARACTERISTICS Insufficient information.
AGE Middle Jurassic, Callovian.
DISTRIBUTION AND FORMATION/S Kyrgyzstan; Balabansai.
NOTES The claim that there are two hand claws is problematic.

Lapparentosaurus madagascariensis
—— Adult size uncertain
FOSSIL REMAINS A few partial skeletons, subadult to juvenile.
ANATOMICAL CHARACTERISTICS Insufficient information.

AGE Middle Jurassic, Bathonian.
DISTRIBUTION AND FORMATION/S Madagascar; Sakahara.
NOTES Used to be considered a brachiosaurid.

Cetiosaurus oxoniensis
—— 16 m (50 ft) TL, 11 tonnes
FOSSIL REMAINS Majority of skeleton.
ANATOMICAL CHARACTERISTICS Neck moderately long.
AGE Middle Jurassic, middle Bathonian.
DISTRIBUTION AND FORMATION/S Central England; Forest Marble.
NOTES First known sauropod.

Cetiosauriscus stewarti
—— 15 m (50 ft) TL, 10 tonnes
FOSSIL REMAINS Partial skeleton.
ANATOMICAL CHARACTERISTICS Insufficient information.
AGE Middle Jurassic, Callovian.
DISTRIBUTION AND FORMATION/S Eastern England; Lower Oxford Clay.
NOTES Not a diplodocoid as was suggested.

Haplocanthosaurus delfsi
—— 16 m (55 ft) TL, 14.4 tonnes
ANATOMICAL CHARACTERISTICS Insufficient information.
AGE Late Jurassic, late Oxfordian.
DISTRIBUTION AND FORMATION/S Colorado; lower Morrison.

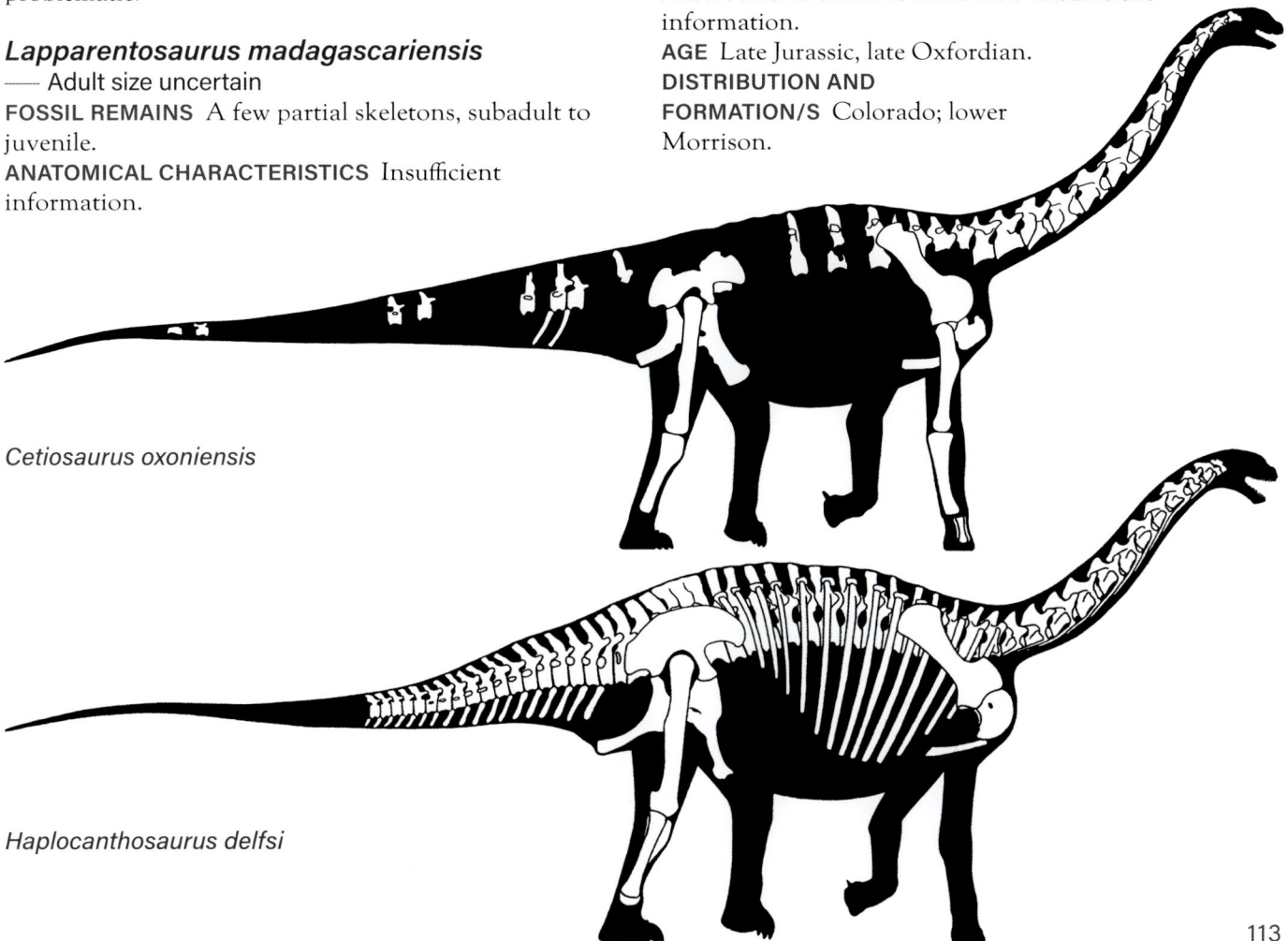

Cetiosaurus oxoniensis

Haplocanthosaurus delfsi

HABITAT Short wet season, otherwise semiarid with open floodplain prairies and riverine forests.
HABITS Probably a feeding generalist.
NOTES Whether this is a different species from the slightly later *H. priscus* is uncertain. *Haplocanthosaurus* may have been a basal diplodocoid.

Haplocanthosaurus priscus
—— 12 m (40 ft) TL, 5.6 tonnes
FOSSIL REMAINS Majority of two skeletons.
ANATOMICAL CHARACTERISTICS Neck moderately long.
AGE Late Jurassic, early and/or early middle Kimmeridgian.
DISTRIBUTION AND FORMATION/S Colorado, Wyoming; lower middle Morrison.

HABITAT Short wet season, otherwise semiarid with open floodplain prairies and riverine forests.
HABITS Probably a feeding generalist.
NOTES May be the ancestor of *H.* unnamed species.

Haplocanthosaurus? unnamed species
—— 12 m (40 ft) TL, 5.5 tonnes
FOSSIL REMAINS Minority of skeleton.
ANATOMICAL CHARACTERISTICS Neck moderately long.
AGE Late Jurassic, early and/or early middle Kimmeridgian.
DISTRIBUTION AND FORMATION/S Colorado; upper middle Morrison.
HABITAT Short wet season, otherwise semiarid with open floodplain prairies and riverine forests.
HABITS Probably a feeding generalist.

MAMENCHISAURS
LARGE TO GIGANTIC EUSAUROPODS LIMITED TO THE MIDDLE AND LATE JURASSIC OF ASIA.

ANATOMICAL CHARACTERISTICS Variable. Heads short, snout rounded. Necks long to extremely long, able to elevate vertically. Tails moderately long. Arms long, so shoulders somewhat higher than hip. Retroverted pelves facilitated slow walking when rearing up by keeping hips and tail horizontal when bipedal, sled-shaped chevrons under tails facilitated static rearing with tails as a prop.
HABITS High-level browsers, both when quadrupedal and when bipedal or tripodal.
NOTES Representing an apparent radiation of Asian sauropods when the continent was isolated; the contents of the group and the relationships of these taxa are obscure, with generic designations often problematic; group probably splittable into a number of divisions. Very incomplete, Late Jurassic East African *Wamweracaudia keranje* may be a non-Asian member of this group, which may have survived into very Early Cretaceous.

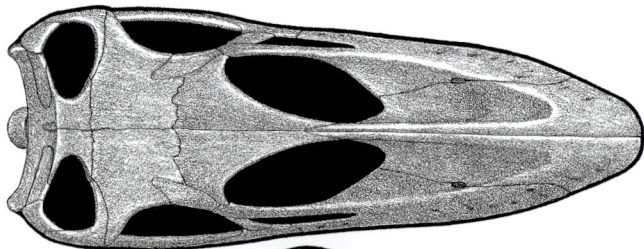

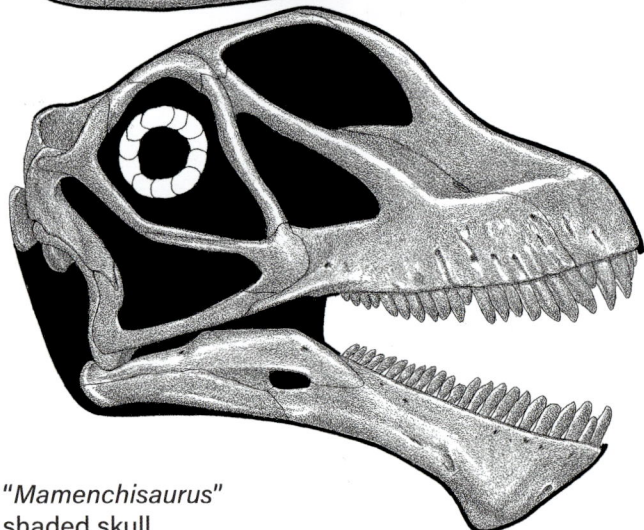

"Mamenchisaurus" shaded skull

Tonganosaurus hei
—— Adult size uncertain
FOSSIL REMAINS Minority of skeleton.
ANATOMICAL CHARACTERISTICS Neck long.
AGE Early Jurassic.
DISTRIBUTION AND FORMATION/S Southern China; Yimin.
NOTES Whether this early sauropod is a mamenchisaur is uncertain.

Chuanjiesaurus anaensis
—— 17 m (55 ft) TL, 11 tonnes
FOSSIL REMAINS Two partial skeletons.
ANATOMICAL CHARACTERISTICS Neck very long. Shoulders a little higher than hips. Tail not large.
AGE Middle Jurassic, Bajocian.
DISTRIBUTION AND FORMATION/S Southwestern China; lower Chuanjie.
NOTES One skeleton may be a sex of this species, or *C.* (or *Analong*) *chuanjieensis*.

Omeisaurus junghsiensis
—— 14 m (45 ft) TL, 4 tonnes
FOSSIL REMAINS Partial skull and skeletons.

ANATOMICAL CHARACTERISTICS Neck very long.
AGE Middle Jurassic, probably Bathonian.
DISTRIBUTION AND FORMATION/S Central China; lower Xiashaximiao.
HABITAT Heavily forested.

Unnamed genus *tianfuensis*

—— 18 m (60 ft) TL, 9.3 tonnes
FOSSIL REMAINS Majority of skull and skeletons.
ANATOMICAL CHARACTERISTICS Neck extremely long and slender. Tail may be tipped by small, roundish club.

Unnamed genus *tianfuensis*

Unnamed
genus
tianfuensis

AGE Middle Jurassic, probably Bathonian.
DISTRIBUTION AND FORMATION/S Central China; lower Xiashaximiao.
HABITAT Heavily forested.
NOTES Too different to be placed in shorter, stouter-necked *Omeisaurus*.

Unnamed genus *jiaoi*
—— 19 m (62 ft) TL, 12 tonnes
FOSSIL REMAINS Majority of skeletons.
ANATOMICAL CHARACTERISTICS Insufficient information.
AGE Middle Jurassic, probably Bathonian.
DISTRIBUTION AND FORMATION/S Central China; lower Xiashaximiao.
HABITAT Heavily forested.
NOTES Too different to be placed in *Omeisaurus* or unnamed genus *tianfuensis*.

Huangshanlong anhuiensis
—— 18 m (60 ft) TL, 9.3 tonnes
FOSSIL REMAINS Minority of one or two skeleton/s.
ANATOMICAL CHARACTERISTICS Insufficient information.
AGE Middle Jurassic.
DISTRIBUTION AND FORMATION/S Eastern China; lower Hongqin.
NOTES May include *Anhuilong diboensis*.

Yuanmousaurus jiangyiensis
—— 17 m (60 ft) TL, 12 tonnes
FOSSIL REMAINS Partial skeleton(s).
ANATOMICAL CHARACTERISTICS Neck long.
AGE Middle Jurassic.
DISTRIBUTION AND FORMATION/S Southern China; Zhanghe.
NOTES *Eomamenchisaurus yuanmouensis* may be a juvenile of this species.

Xinjiangtitan shanshanesis
—— 34 m (110 ft) TL, 40 tonnes
FOSSIL REMAINS Majority of skeleton.
ANATOMICAL CHARACTERISTICS Neck extremely long, both absolutely and relative to body.
AGE Middle Jurassic, Callovian.

DISTRIBUTION AND FORMATION/S Northwestern China; Qiketai.
HABITS Could feed on tree crowns 20 m high (65 ft) quadrupedally, 23 m high (75 ft) bipedally or tripodally.
NOTES Earliest known gigantic sauropod, dinosaur, and land animal, 16 m long (52 ft) neck longest known, especially based on a complete vertebral series.

Klamelisaurus gobiensis
—— 13 m (45 ft) TL, 8.2 tonnes
FOSSIL REMAINS Majority of skeleton.
ANATOMICAL CHARACTERISTICS Neck long. Shoulders a little higher than hips. Tail not large.
AGE Middle Jurassic, late Callovian.
DISTRIBUTION AND FORMATION/S Northwestern China; lower Shishugou.

Tienshanosaurus chitaiensis
—— Adult size uncertain
FOSSIL REMAINS partial skeleton, possibly immature.
ANATOMICAL CHARACTERISTICS Insufficient skeleton.
AGE Late Jurassic, early? Oxfordian.
DISTRIBUTION AND FORMATION/S Northwestern China; upper Shishugou.

Rhomaleopakhus turpanensis
—— 25 m (80 ft) TL, 25 tonnes
FOSSIL REMAINS Minority of skeleton.
ANATOMICAL CHARACTERISTICS Forelimb robust, thumb claw large.
AGE Late Jurassic, upper Kimmeridgian or Tithonian.
DISTRIBUTION AND FORMATION/S Northwestern China; lower Kalazha.
NOTES Probably not the same as fragmentary *Hudiesaurus sinojapanorum*.

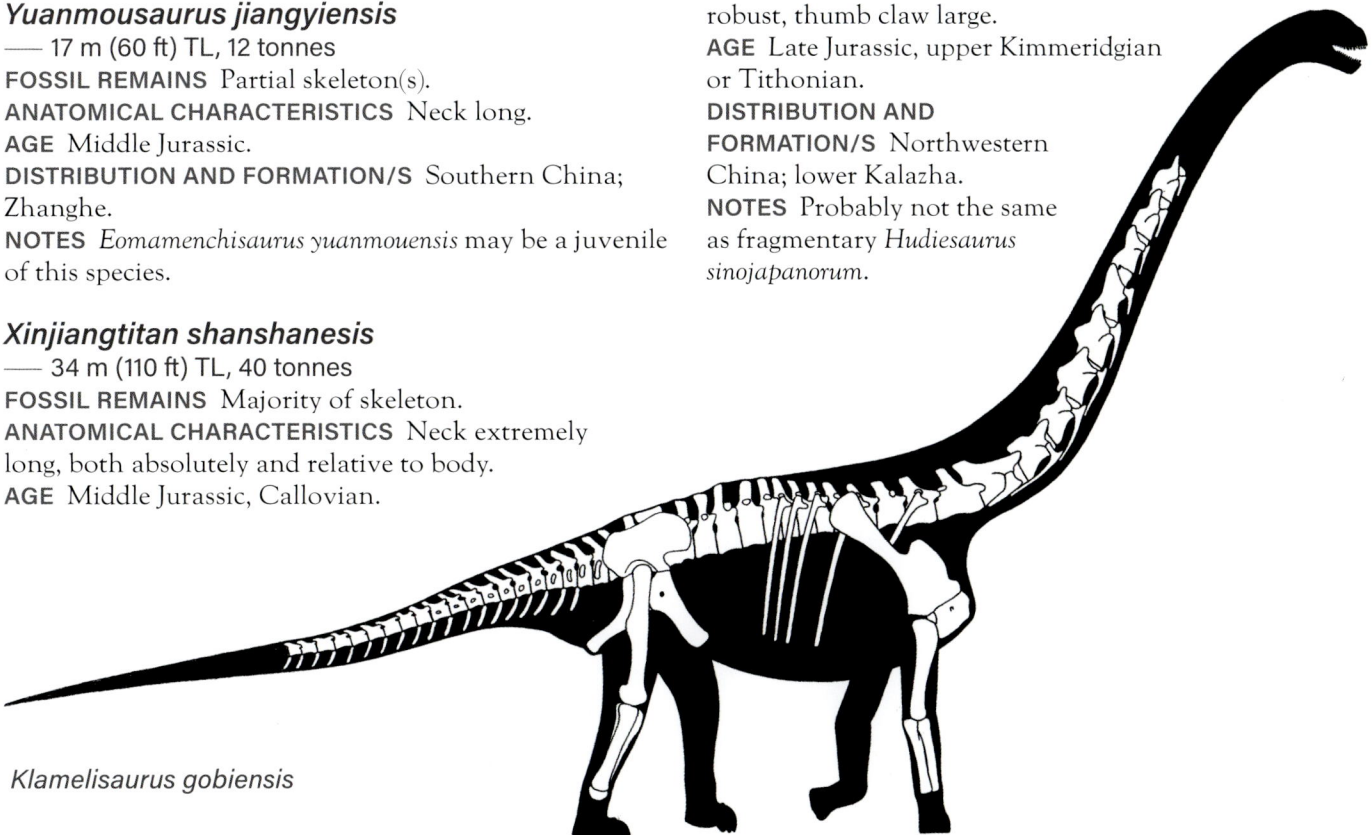

Klamelisaurus gobiensis

Unnamed genus *sinocanadorum*

—— Adult size uncertain

FOSSIL REMAINS Minority of skull and minority of skeleton(s), possible juveniles.

ANATOMICAL CHARACTERISTICS Neck extremely long.

AGE Late Jurassic, early? Oxfordian.

DISTRIBUTION AND FORMATION/S Northwestern China; upper Shishugou.

NOTES A pair of neck vertebrae that may belong to this species are about as long as those of *Xinjiangtitan*, may indicate an individual of same size, or up to ~90 tonnes if proportions were similar to unnamed genus *hochuanensis*, in which case is one of largest known land animals. *Bellusaurus sui* may be juvenile of this species.

Tongnanlong? *zhimingi*

—— 25 m (80 ft) TL, 25 tonnes

FOSSIL REMAINS Minority of skeleton.

ANATOMICAL CHARACTERISTICS Insufficient information.

AGE Late Jurassic.

DISTRIBUTION AND FORMATION/S Central China; lower Suining.

NOTES May belong to an already named genus, or this genus name may apply to other mamenchisaurs such as the next two species.

Unnamed genus *hochuanensis*

—— 21 m (70 ft) TL, 15.5 tonnes

FOSSIL REMAINS Partial skull and a few skeletons.

ANATOMICAL CHARACTERISTICS Neck extremely long, vertebral spines near base of neck forked. Tail tipped by small, roundish club. Limbs short.

AGE Late Jurassic, probably Oxfordian.

DISTRIBUTION AND FORMATION/S Central China; lower Shangshaximiao.

HABITAT Heavily forested.

HABITS Purpose of very small tail club uncertain.

NOTES Too different from shorter-necked *Mamenchisaurus* to be in same genus. Placement of some fossils in this species uncertain.

Unnamed genus *youngi*?

—— 17 m (55 ft) TL, 7.7 tonnes

FOSSIL REMAINS Complete skull and majority of skeleton.

ANATOMICAL CHARACTERISTICS Neck extremely long, vertebral spines near base of neck forked. Hip strongly retroverted, and tail directed strongly upward. Limbs short.

AGE Late Jurassic, probably Oxfordian.

DISTRIBUTION AND FORMATION/S Central China; lower Shangshaximiao.

HABITAT Heavily forested.

NOTES Among the most peculiarly shaped sauropods. May be a sex of unnamed genus *hochuanensis*.

Mamenchisaurus constructus

—— 15 m (50 ft) TL, 5 tonnes

FOSSIL REMAINS Minority of skeleton.

ANATOMICAL CHARACTERISTICS Neck moderately long.

AGE Late Jurassic, probably Oxfordian.

DISTRIBUTION AND FORMATION/S Central China; lower Shangshaximiao.

HABITAT Heavily forested.

NOTES Based on an inadequate fossil without a very long neck, that so many species have been placed in *Mamenchisaurus*, many from the same formation, indicates that these sauropods are overlumped, being in the wrong genus in some cases, or split into too many species in others. *Daanosaurus zhangi* may be a juvenile of one of the mamenchisaurids from the Shangshaximiao Formation.

Unnamed genus *maoianus*

—— 15 m (50 ft) TL, 5 tonnes

FOSSIL REMAINS Nearly complete skull and partial skeleton.

ANATOMICAL CHARACTERISTICS Neck very long.

AGE Late Jurassic, probably Oxfordian.

DISTRIBUTION AND FORMATION/S Central China; lower Shangshaximiao.

HABITAT Heavily forested.

Unnamed genus *maoianus*

Uncertain genus *jingyanensis*?

—— 20 m (65 ft) TL, 13 tonnes

FOSSIL REMAINS Majority of skull and minority of skeleton.

ANATOMICAL CHARACTERISTICS Neck extremely long.

AGE Late Jurassic, probably Oxfordian.

DISTRIBUTION AND FORMATION/S Central China; lower Shangshaximiao.

HABITAT Heavily forested.

NOTES Probably belongs to one of the other incomplete species from the lower Shangshaximiao, possibly *Mamenchisaurus*.

uncertain genus *jingyanensis*?

Jingiella dongxingensis
—— 15 m (50 ft) TL, 5 tonnes
FOSSIL REMAINS Minority of skeleton.
ANATOMICAL CHARACTERISTICS Insufficient
information.
AGE Late Jurassic, probably Kimmeridgian.
DISTRIBUTION AND FORMATION/S Southern China;
middle Dongxing.

Qijianglong guokr
—— Adult size uncertain
FOSSIL REMAINS Partial skeleton, probably large
juvenile.
ANATOMICAL CHARACTERISTICS Insufficient
information.
AGE Early Cretaceous, late Aptian.
DISTRIBUTION AND FORMATION/S
Central China; Suining.

Unnamed genus *hochuanensis*

Unnamed genus *youngi*?

Unnamed genus *anyuensis*
—— 25 m (80 ft) TL, 25 tonnes
FOSSIL REMAINS Several partial skeletons.
ANATOMICAL CHARACTERISTICS Neck extremely long.
AGE Early Cretaceous, late Aptian.

DISTRIBUTION AND FORMATION/S Central China; Suining.
NOTES Cannot be in much earlier *Mamenchisaurus*, indicates group survived well into Cretaceous.

TURIASAURS
MEDIUM-SIZED TO ENORMOUS SAUROPODS OF THE LATE JURASSIC AND EARLY CRETACEOUS.

ANATOMICAL CHARACTERISTICS Necks and tails moderately long. Arms moderately long, so shoulders about as high as hips.
NOTES Turisaurs are proving to be more widely distributed than first realized.

Narindasaurus thevenini
—— 10 m (30 ft) TL, 2 tonnes
FOSSIL REMAINS Minority of skeleton.
ANATOMICAL CHARACTERISTICS Insufficient information.
AGE Middle Jurassic, Bathonian.
DISTRIBUTION AND FORMATION/S Madagascar; middle Isalo III.

Amanzia greppini
—— 10 m (30 ft) TL, 2 tonnes
FOSSIL REMAINS Partial remains.
ANATOMICAL CHARACTERISTICS Insufficient information.
AGE Late Jurassic, early Kimmeridgian.
DISTRIBUTION AND FORMATION/S Switzerland; lower Reuchenette.

Tendaguria tanzaniensis
—— 20 m (65 ft) TL, 15 tonnes
FOSSIL REMAINS Minority of skeleton.
ANATOMICAL CHARACTERISTICS Insufficient information.
AGE Late Jurassic, late Kimmeridgian and/or early Tithonian?.
DISTRIBUTION AND FORMATION/S Tanzania; upper? Tendaguru.
HABITAT Coastal, seasonally dry with heavier vegetation inland.

Turiasaurus riodevensis
—— 30 m (100 ft) TL, 50 tonnes
FOSSIL REMAINS Partial skeletons.
ANATOMICAL CHARACTERISTICS Some neck and trunk vertebral spines forked.
AGE Late Jurassic and/or early Cretaceous, late Tithonian and/or earliest Berriasian.
DISTRIBUTION AND FORMATION/S Eastern Spain; Villar del Arzobispo.
NOTES The largest known nonneosauropod.

Zby atlanticus
—— 17 m (55 ft) TL, 12 tonnes
FOSSIL REMAINS Minority of skeleton.
ANATOMICAL CHARACTERISTICS Insufficient information.
AGE Late Jurassic, early or middle Tithonian.
DISTRIBUTION AND FORMATION/S Portugal; middle Lourinhã.
HABITAT Large, seasonally dry island with open woodlands.

Losillasaurus giganteus
—— Adult size uncertain
FOSSIL REMAINS Minority of several skeletons.
ANATOMICAL CHARACTERISTICS Vertebral spines not forked.
AGE Late Jurassic and/or early Cretaceous, late Tithonian and/or earliest Berriasian.
DISTRIBUTION AND FORMATION/S Eastern Spain; upper Villar del Arzobispo.
NOTES Subadult remains indicate a very large sauropod.

Mierasaurus bobyoungi
—— Adult size uncertain
FOSSIL REMAINS Partial skull and skeleton, other remains of varying ages.
ANATOMICAL CHARACTERISTICS Vertebral spines not forked.
AGE Early Cretaceous, late Berriasian.
DISTRIBUTION AND FORMATION/S Utah; lowermost Cedar Mountain.

Moabosaurus utahensis
—— Adult size uncertain
FOSSIL REMAINS Disarticulated bone bed remains of varying ages.
ANATOMICAL CHARACTERISTICS Vertebral spines not forked.
AGE Early Cretaceous, late Valanginian.
DISTRIBUTION AND FORMATION/S Utah; lower Cedar Mountain.

NEOSAUROPODS

LARGE TO ENORMOUS EUSAUROPODS OF THE EARLY JURASSIC TO THE END OF THE DINOSAUR ERA, MOST CONTINENTS.

ANATOMICAL CHARACTERISTICS Skeletal pneumaticity and birdlike respiratory system well developed.
NOTES Absence from Antarctica probably reflects lack of sufficient sampling.

DIPLODOCOIDS

SMALL (FOR SAUROPODS) TO GIGANTIC NEOSAUROPODS LIMITED TO THE EARLY JURASSIC TO THE EARLY LATE CRETACEOUS OF THE AMERICAS, EUROPE, AFRICA, AND ASIA.

ANATOMICAL CHARACTERISTICS Variable. Heads long, shallow, bony nostrils strongly retracted to above the orbits, but fleshy nostrils probably still near front of snout, which is broad and squared off. Shallow lower jaws short, pencil-shaped teeth limited to front of jaws, head flexed downward relative to neck. Necks short to extremely long, not carried strongly erect. Very long tails ending in a whip, tip may have been able to achieve supersonic speeds. Arms and hands short, so shoulders lower than hips, which are heightened by tall vertebral spines. Short arms, large hips, and heavy tails with sled-shaped chevrons facilitated static rearing posture.
HABITS Flexible feeders able to easily browse and graze at all levels from the ground up to very high. High rates of grit wear on teeth occurred when grazing, worn-down teeth were rapidly replaced.
NOTES Absence from Australia and Antarctica probably reflects lack of sufficient sampling.

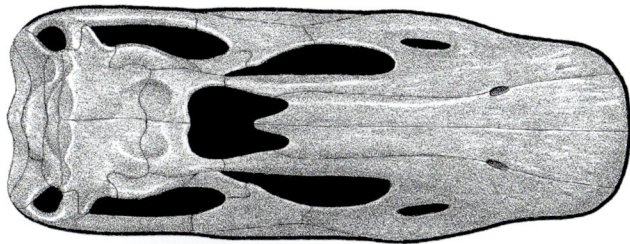

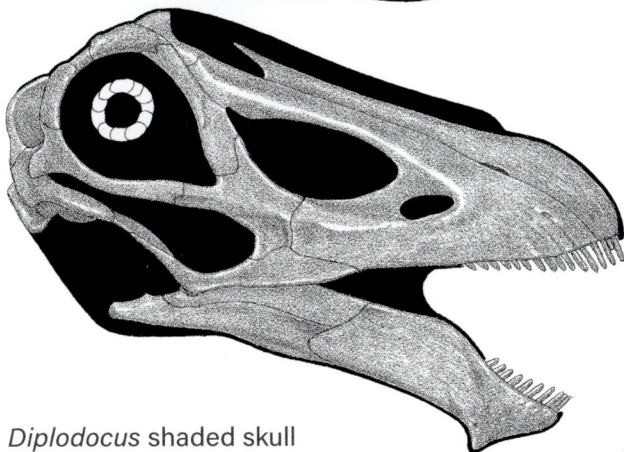

Diplodocus shaded skull

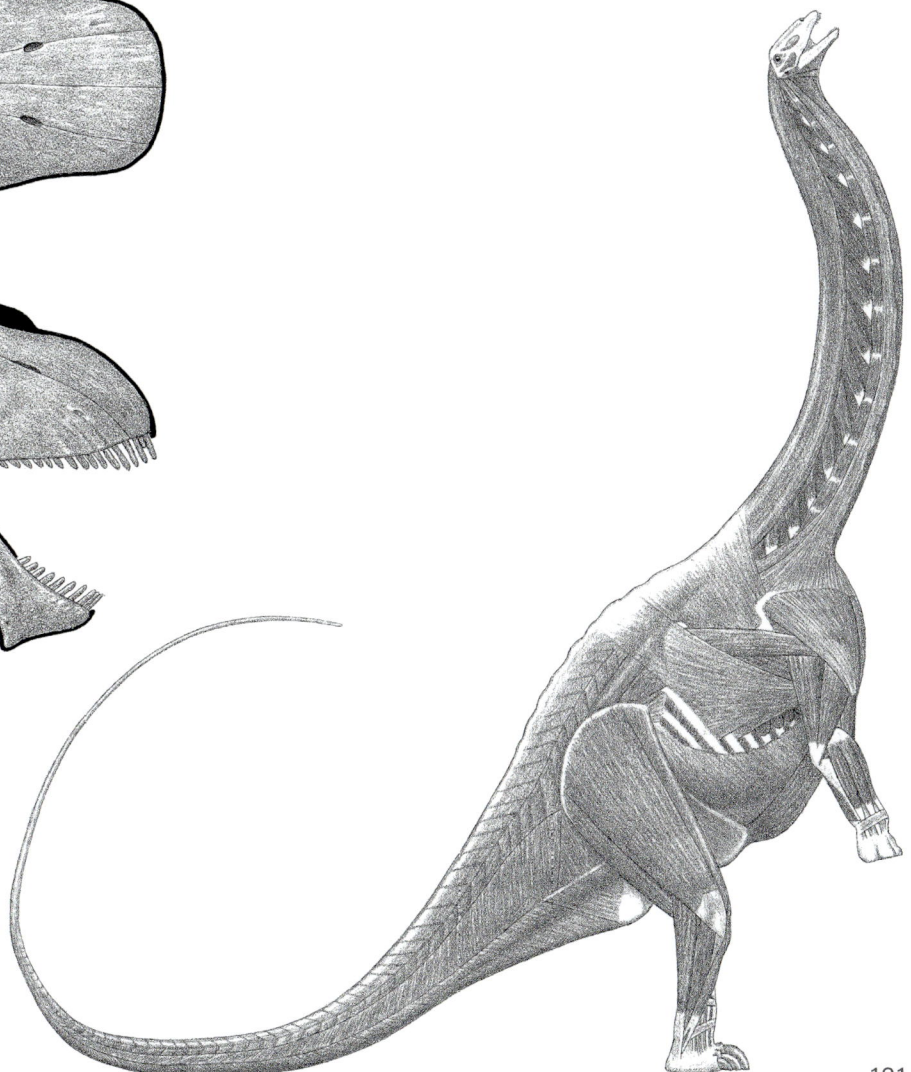

Brontosaurus muscle study

121

REBBACHISAURIDS

SMALL AND MEDIUM-SIZED DIPLODOCOIDS LIMITED TO THE EARLY AND EARLY LATE CRETACEOUS OF SOUTH AMERICA AND AFRICA.

ANATOMICAL CHARACTERISTICS Fairly uniform. Necks short by sauropod standards, neck ribs overlapping a little. Vertebral spines not forked. Upper scapula blades very broad.

NOTES The last radiation of diplodocids and nonmacronarian sauropods. Fragmentary *Xenoposeidon proneneukos* from the Berriasian-Valanginian of England may be the earliest known example of this group. Early Late Cretaceous examples were the only nontitanosaur sauropods of the Late Cretaceous. A fossil labeled *Maraapunisaurus* (was *Amphicoelias*) *fragillimus*—based on a disintegrated colossal trunk vertebra up to 2.6 m tall (8.5 ft) from the Late Jurassic upper Morrison Formation—may be related to this group, and suggests animals of 80–120 tonnes.

Amazonsaurus maranhensis
—— 12 m (40 ft) TL, 4.5 tonnes
FOSSIL REMAINS Minority of skeleton.
ANATOMICAL CHARACTERISTICS Insufficient information.
AGE Early Cretaceous, Aptian or Albian.
DISTRIBUTION AND FORMATION/S Northern Brazil; Itapecuru.
NOTES Relationships of *Amazonsaurus* are uncertain.

Zapalasaurus bonapartei
—— 9 m (29 ft) TL, 2 tonnes
FOSSIL REMAINS Partial skeleton.
ANATOMICAL CHARACTERISTICS Insufficient information.
AGE Early Cretaceous, late Barremian.
DISTRIBUTION AND FORMATION/S Western Argentina; upper La Amarga.
HABITAT Well-watered coastal woodlands with short dry season.

Agustina ligabuei
—— 15 m (50 ft) TL, 8 tonnes
FOSSIL REMAINS Minority of skeleton.
ANATOMICAL CHARACTERISTICS Insufficient information.
AGE Early Cretaceous, late Aptian and/or early Albian.
DISTRIBUTION AND FORMATION/S Western Argentina; Lohan Cura.
HABITAT Well-watered coastal woodlands with short dry season.
NOTES Not armored with spikes and plates as originally thought.

Comahuesaurus windhauseni
—— 12 m (40 ft) TL, 4.5 tonnes
FOSSIL REMAINS Partial skeleton.
ANATOMICAL CHARACTERISTICS Insufficient information.
AGE Early Cretaceous, late Aptian and/or early Albian.
DISTRIBUTION AND FORMATION/S Western Argentina; Lohan Cura.
HABITAT Well-watered coastal woodlands with short dry season.

Lavocatisaurus agrioensis
—— 11 m (35 ft) TL, 3 tonnes
FOSSIL REMAINS Partial adult skull and skeleton and two partial juvenile skeletons.
ANATOMICAL CHARACTERISTICS Snout broad and squared off.
AGE Early Cretaceous, late Aptian or early Albian.
DISTRIBUTION AND FORMATION/S Western Argentina; upper Rayoso.

Katepensaurus goicoecheal
—— Adult size uncertain
FOSSIL REMAINS Partial skull and skeleton, immature.
ANATOMICAL CHARACTERISTICS Standard for group.
AGE Late Cretaceous, late Cenomanian or Turonian.
DISTRIBUTION AND FORMATION/S Southern Argentina; lower Bajo Barreal.
HABITAT Seasonally wet, well-forested floodplain.

Sidersaura marae
—— 17 m (55 ft) TL, 12 tonnes
FOSSIL REMAINS Partial skeletons, one juvenile.
ANATOMICAL CHARACTERISTICS Standard for group.
AGE Late Cretaceous, upper Cenomanian or Turonian.
DISTRIBUTION AND FORMATION/S Western Argentina; lowest Huincul.
HABITAT Seasonally arid open woodlands.

Cienciargentina sanchezi
—— Adult size uncertain
FOSSIL REMAINS Three partial skeletons, varying sizes.
ANATOMICAL CHARACTERISTICS Standard for group.
AGE Late Cretaceous, Cenomanian or Turonian.
DISTRIBUTION AND FORMATION/S Western Argentina; lower Huincul.
HABITAT Seasonally arid open woodlands.

REBBACHISAURIDS ◄ **SAUROPODS**

Cathartesaura anaerobica
—— 12 m (40 ft) TL, 3 tonnes
FOSSIL REMAINS Partial skeleton.
ANATOMICAL CHARACTERISTICS Insufficient information.
AGE Late Cretaceous, middle Cenomanian.
DISTRIBUTION AND FORMATION/S Western Argentina; lower Huincul.
HABITAT Seasonally arid open woodlands.
NOTES This and *Rebbachisaurus* species may form limaysaurines subfamily.

Rebbachisaurus (or Rayosaurus) agrioensis
—— 10 m (33 ft) TL, 2.5 tonnes
FOSSIL REMAINS Minority of skeleton.
ANATOMICAL CHARACTERISTICS Insufficient information.
AGE Early Cretaceous, Aptian.
DISTRIBUTION AND FORMATION/S Western Argentina; Rayoso.
NOTES Whether *Rayosaurus*, *Rebbachisaurus*, and *Limaysaurus* are separate genera is uncertain.

Limaysaurus (or Rebbachisaurus) tessonei
—— 15 m (50 ft) TL, 8 tonnes
FOSSIL REMAINS Minority of skull, majority of skeleton.
ANATOMICAL CHARACTERISTICS Neck fairly deep. Tall vertebral spines over hips form a low sail. Chevrons may be absent from most of underside of tail.
AGE Early Cretaceous, early Cenomanian.
DISTRIBUTION AND FORMATION/S Western Argentina; lower Candeleros.
HABITAT Short wet season, otherwise semiarid with open floodplains and riverine forests.

Rebbachisaurus garasbae
—— 14 m (45 ft) TL, 7.4 tonnes
FOSSIL REMAINS Partial skeleton.
ANATOMICAL CHARACTERISTICS Hip sail tall.
AGE Early Cretaceous, Albian.
DISTRIBUTION AND FORMATION/S Morocco; Tegana.

Demandasaurus darwini
—— 9 m (30 ft) TL, 2 tonnes
FOSSIL REMAINS Minority of skull and skeletons, many isolated bones.
ANATOMICAL CHARACTERISTICS Neck short. Hip sail present.
AGE Early Cretaceous, late Barremian or early Aptian.
DISTRIBUTION AND FORMATION/S Spain; Castrillo de la Reina.
NOTES This, *Itapeuasaurus*, *Tataouinea*, *Nigersaurus* may form subfamily Nigersaurinae.

Tataouinea hannibalis
—— 14 m (45 ft) TL, 8 tonnes
FOSSIL REMAINS Minority of skeleton.
ANATOMICAL CHARACTERISTICS Hip sail present.
AGE Early Cretaceous, early Albian.
DISTRIBUTION AND FORMATION/S Tunisia; Aïn el Guettar.
HABITAT Coastal.

Nigersaurus taqueti
—— 9 m (30 ft) TL, 2 tonnes
FOSSIL REMAINS Majority of skull, several partial skeletons, many isolated bones.

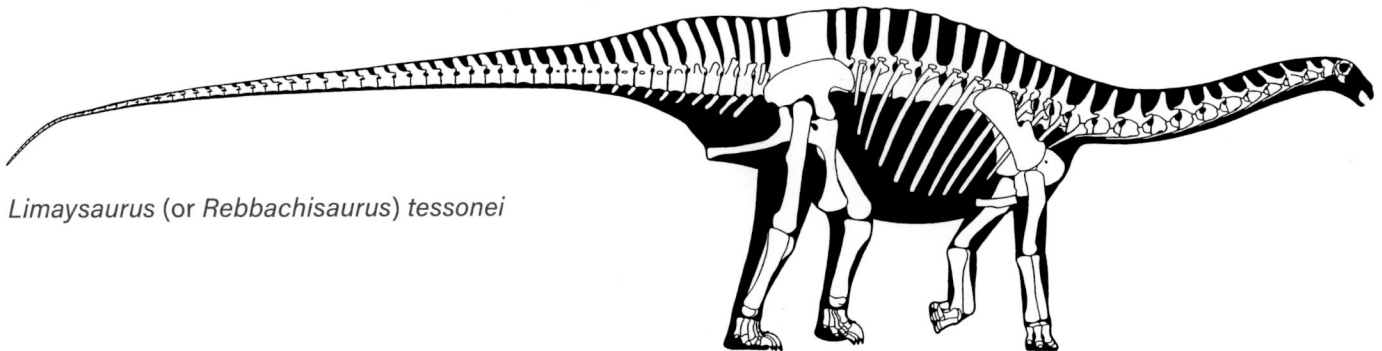

Limaysaurus (or Rebbachisaurus) tessonei

Nigersaurus taqueti
(see also next page)

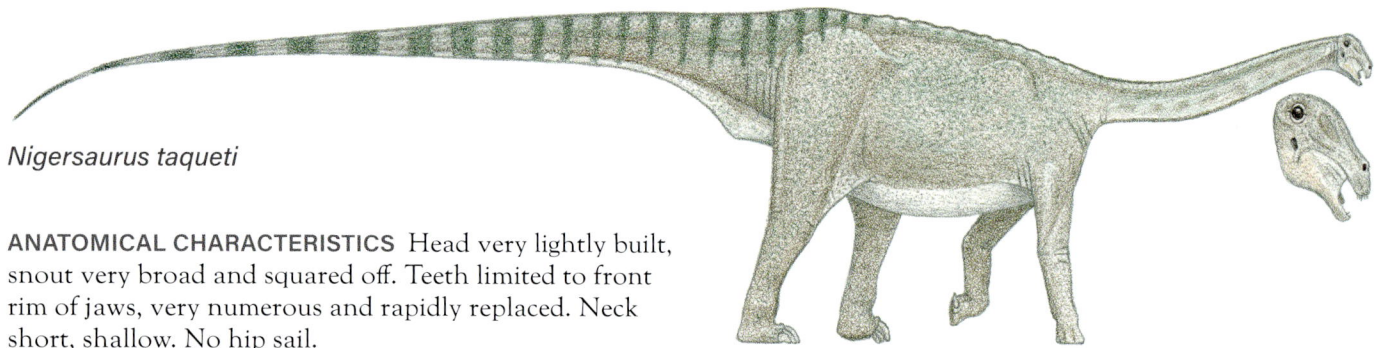

Nigersaurus taqueti

ANATOMICAL CHARACTERISTICS Head very lightly built, snout very broad and squared off. Teeth limited to front rim of jaws, very numerous and rapidly replaced. Neck short, shallow. No hip sail.
AGE Early Cretaceous, late Aptian.
DISTRIBUTION AND FORMATION/S Niger; upper Elrhaz.
HABITAT Coastal river delta.
HABITS Square muzzle at end of long neck was adaptation for mowing ground cover, also able to rear to high browse.
NOTES The most complex tooth battery among saurischians, mimics in some regards battery of ornithischians except teeth were only for cropping plants. It is not known how many other rebbachisaurids shared these feeding adaptations. The other sauropod known to have a similarly broad and square beak is the titanosaur *Bonitasaura.*

Itapeuasaurus cajapioensis
— 9 m (30 ft) TL, 2 tonnes
FOSSIL REMAINS Minority of skeleton.
ANATOMICAL CHARACTERISTICS Insufficient information.
AGE Late Cretaceous, Cenomanian.
DISTRIBUTION AND FORMATION/S Northeastern Brazil; Alcantara.

Nigersaurus taqueti

DICRAEOSAURIDS

SMALL (BY SAUROPOD STANDARDS) DIPLODOCOIDS LIMITED TO THE EARLY JURASSIC TO THE EARLY CRETACEOUS OF SOUTH AMERICA, AFRICA, AND ASIA.

ANATOMICAL CHARACTERISTICS Uniform. Necks short by sauropod standards, spines usually very tall, unable to elevate above shoulder level, ribs so short they do not overlap, increasing flexibility of neck. Tall vertebral spines over hips form a low sail. Most neck and trunk vertebral spines forked.
NOTES Absence from other continents may reflect lack of sufficient sampling.

Tharosaurus indicus
—— 10 m (35 ft) TL, 2 tonnes
FOSSIL REMAINS Minority of skeleton.
ANATOMICAL CHARACTERISTICS Insufficient information.
AGE Middle Jurassic, early middle Bathonian.
DISTRIBUTION AND FORMATION/S Northern India; lower Jaisalmer.
NOTES Oldest known diplodocoid.

Lingwulong shenqi
—— Adult size uncertain
FOSSIL REMAINS Minority of skull, partial skeleton, other partial remains, probably immature.
ANATOMICAL CHARACTERISTICS Neck spines not tall.
AGE Middle or Late Jurassic.
DISTRIBUTION AND FORMATION/S Northeastern China; uncertain.
NOTES Had been thought to be the oldest known diplodocoid.

Suuwassea emilieae
—— 15 m (50 ft) TL, 6 tonnes
FOSSIL REMAINS Minority of skull and skeleton.
ANATOMICAL CHARACTERISTICS Neck spines not tall.
AGE Late Jurassic, middle? Kimmeridgian.

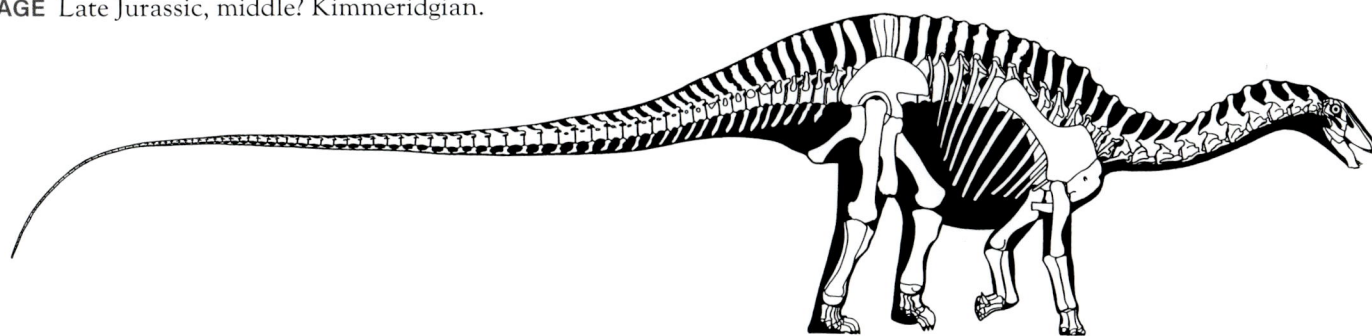

DISTRIBUTION AND FORMATION/S Montana; probably middle Morrison.
HABITAT More coastal and wetter than rest of Morrison.
NOTES Relationships of this diplodocoid are uncertain.

Brachytrachelopan mesai
—— 10 m (35 ft) TL, 2 tonnes
FOSSIL REMAINS Partial skeleton.
ANATOMICAL CHARACTERISTICS Neck short, spines not tall.
AGE Early Jurassic, middle Toarcian.
DISTRIBUTION AND FORMATION/S Southern Argentina; upper Cañadón Asfalto.
NOTES The shortest-necked known sauropod.

Dicraeosaurus hansemanni
—— 14 m (45 ft) TL, 5.5 tonnes
FOSSIL REMAINS Minority of skull, several skeletons from nearly complete to partial.
ANATOMICAL CHARACTERISTICS Standard for group, including lower jaw, did not have the contorted lower edge with which it is usually restored.
AGE Late Jurassic, middle Kimmeridgian.

Dicraeosaurus hansemanni

125

DISTRIBUTION AND FORMATION/S Tanzania; middle Tendaguru.

HABITAT Coastal, seasonally dry with heavier vegetation inland.

NOTES Remains from much earlier lower Tendaguru are not same taxon. May be an ancestor of *D. sattleri*.

Dicraeosaurus? sattleri
—— 15 m (50 ft) TL, 7 tonnes

FOSSIL REMAINS Minority of skull, several partial skeletons.

ANATOMICAL CHARACTERISTICS Probably similar to *D. hansemanni*.

AGE Late Jurassic, late Kimmeridgian and/or early Tithonian.

DISTRIBUTION AND FORMATION/S Tanzania; upper Tendaguru.

HABITAT Coastal, seasonally dry with heavier vegetation inland.

NOTES May not be the same genus as much earlier *D. hansemanni*.

Smitanosarus agilis
—— Adult size uncertain

FOSSIL REMAINS Minority of skull and skeleton, probably immature.

ANATOMICAL CHARACTERISTICS Insufficient information.

AGE Late Jurassic, early and/or early middle Kimmeridgian.

DISTRIBUTION AND FORMATION/S Colorado; middle Morrison.

HABITAT Short wet season, otherwise semiarid with open floodplain prairies and riverine forests.

Bajadasaurus pronuspinax
—— 10 m (35 ft) TL, 2 tonnes

FOSSIL REMAINS Partial skull and minority of skeleton.

ANATOMICAL CHARACTERISTICS Neck vertebral spines elongated into very long forward-curved spikes that may have been lengthened by horn sheaths.

AGE Early Cretaceous, late Barremian or Valanginian.

DISTRIBUTION AND FORMATION/S Western Argentina; Bajada Colorado.

HABITS Defense included arc of neck spines. Latter may have been used to generate clattering noise display.

Pilmatueia faundezi
—— 10 m (35 ft) TL, 2 tonnes

FOSSIL REMAINS Minority of a few skeletons.

ANATOMICAL CHARACTERISTICS Insufficient informaiton.

AGE Early Cretaceous, Valanginian.

DISTRIBUTION AND FORMATION/S Western Argentina; middle Mulichinco.

Amargasaurus cazaui
—— 13 m (43 ft) TL, 4 tonnes

FOSSIL REMAINS Minority of skull and majority of skeleton.

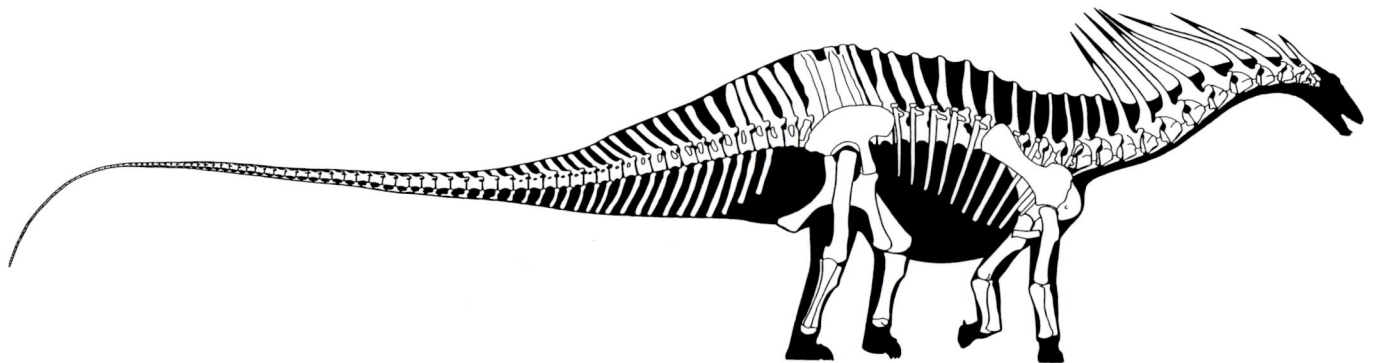

Amargasaurus cazaui

OPPOSITE:
Amargasaurus cazaui

ANATOMICAL CHARACTERISTICS Neck vertebral spines elongated into very long spikes that may have been lengthened by horn sheaths. Hip sail tall.
AGE Late Early Cretaceous, early Barremian.
DISTRIBUTION AND FORMATION/S Western Argentina; lower La Amarga.

HABITAT Well-watered coastal woodlands with short dry season.
HABITS Defense included arc of neck spines. Latter may have been used to generate clattering noise display.
NOTES *Amargatitanis macni* probably adult of this species. Has been suggested that the neck spikes supported sail fins, but this is not likely.

DIPLODOCIDS

LARGE TO GIGANTIC DIPLODOCOIDS LIMITED TO THE MIDDLE JURASSIC TO EARLY CRETACEOUS OF NORTH AMERICA, EUROPE, AND AFRICA.

ANATOMICAL CHARACTERISTICS Variable. Necks long to extremely long, unable to elevate vertically, ribs so short they do not overlap, increasing flexibility of neck. Most neck and trunk vertebral spines forked. Tall vertebral spines over hips form a low sail. Tail whips very long.
NOTES If a diplodocid, fragmentary Argentinian *Leinkupal laticauda* extends group into early Early Cretaceous.

Amphicoelias altus
—— 18 m (60 ft) TL, 5 tonnes
FOSSIL REMAINS Minority of skull and skeleton.
ANATOMICAL CHARACTERISTICS Neck may not be elongated. Leg very slender by sauropod standards.
AGE Late Jurassic, late middle and/or late Kimmeridgian.

DISTRIBUTION AND FORMATION/S Colorado, Wyoming; upper Morrison.
HABITAT Wetter than earlier Morrison, otherwise semiarid with open floodplain prairies and riverine forests.

DIPLODOCINES

LARGE TO GIGANTIC DIPLODOCIDS LIMITED TO THE MIDDLE AND LATE JURASSIC OF NORTH AMERICA, EUROPE, AND AFRICA.

ANATOMICAL CHARACTERISTICS Fairly uniform. Lightly built. Necks very to extremely long, fairly slender. Tails very long. Femurs usually slender. Short vertical spikes appear to run atop vertebral series in at least some diplodocines.
NOTES Incompletely known *Ardetosaurus viator* may be the sexual morph and/or juvenile of one of the diplodocines from the middle Morrison Formation.

Unnamed genus and species
—— 22 m (71 ft) TL, 8.2 tonnes
FOSSIL REMAINS Virtually complete skull and skeleton.
ANATOMICAL CHARACTERISTICS Femur robust.
AGE Late Jurassic, early late Oxfordian.
DISTRIBUTION AND FORMATION/S Wyoming; lowermost Morrison.
HABITAT Short wet season, otherwise semiarid with open floodplain prairies and riverine forests.

Dyslocosaurus polyonychius
—— 18 m (60 ft) TL, 5 tonnes
FOSSIL REMAINS Minority of skeleton.
ANATOMICAL CHARACTERISTICS Insufficient information.
AGE Probably Late Jurassic.
DISTRIBUTION AND FORMATION/S Wyoming; probably Morrison.
HABITAT Short wet season, otherwise semiarid with open floodplain prairies and riverine forests.

NOTES Neither the formation this was found in nor its relationships are entirely certain.

Galeamopus (or Diplodocus) hayi
—— Adult size uncertain
FOSSIL REMAINS Majority of an immature skeleton.
ANATOMICAL CHARACTERISTICS Neck very long. Tail extremely long.
AGE Late Jurassic, late Oxfordian.
DISTRIBUTION AND FORMATION/S Colorado, Wyoming; lower Morrison.
HABITAT Short wet season, otherwise semiarid with open floodplain prairies and riverine forests.
NOTES May be direct ancestor of *G. pabsti*.

Galeamopus (or Diplodocus) pabsti
—— Adult size uncertain
FOSSIL REMAINS Partial skull and skeleton, partial skull.
ANATOMICAL CHARACTERISTICS Neck very long.
AGE Late Jurassic, early and/or early middle Kimmeridgian.

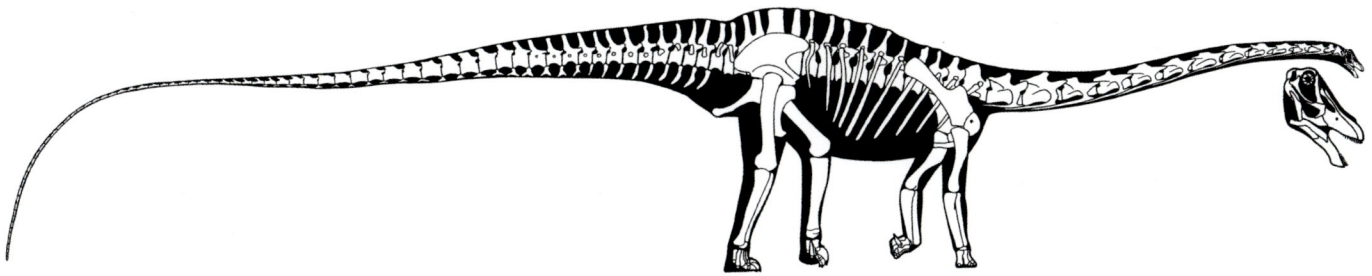

Unnamed genus and species

Galeamopus
(or *Diplodocus*) *pabsti*

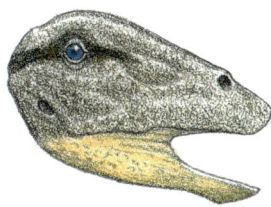

DISTRIBUTION AND FORMATION/S Wyoming, Colorado; middle Morrison.
HABITAT Short wet season, otherwise semiarid with open floodplain prairies and riverine forests.

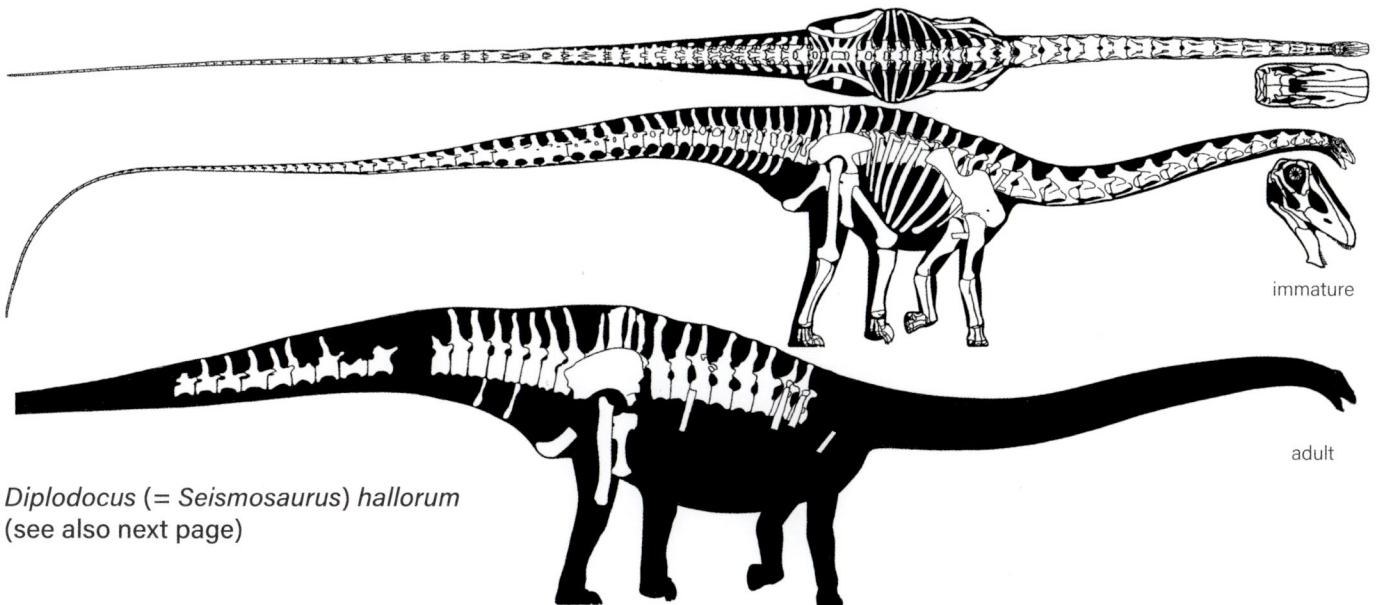

Diplodocus (= *Seismosaurus*) *hallorum*
— 29 m (95 ft) TL, 23 tonnes

FOSSIL REMAINS Possible skull(s), fairly complete and partial skeletons.
ANATOMICAL CHARACTERISTICS Neck very long. Trunk compact. Tail extremely long. Femur slender until mature.
AGE Late Jurassic, early and/or early middle Kimmeridgian.
DISTRIBUTION AND FORMATION/S Colorado, Utah, New Mexico; middle Morrison.
HABITAT Short wet season, otherwise semiarid with open floodplain prairies and riverine forests.
HABITS Broadening of snout and shortening of tooth rows with growth indicates shift to courser vegetation with maturity.

immature

adult

Diplodocus (= *Seismosaurus*) *hallorum*
(see also next page)

NOTES Many remains were placed in *D. longus*, but that is based on very fragmentary remains that may not be the same type as skeletons commonly placed in genus. The size of the largest *"Seismosaurus"* skeleton was greatly exaggerated. It is possible that the 22 m long (72 ft), gracilely legged skeletons from the northern Morrison are a species distinct from the 29 m (95 ft), apparently shorter-limbed *D. hallorum* from the southern Morrison.

Diplodocus carnegii
—— 24 m (80 ft) TL, 12.5 tonnes
FOSSIL REMAINS Majority of several skeletons.
ANATOMICAL CHARACTERISTICS Neck very long. Trunk fairly long. Tail extremely long. Femur slender, at least at known sizes.
AGE Late Jurassic, early and/or early middle Kimmeridgian.

Diplodocus carnegii

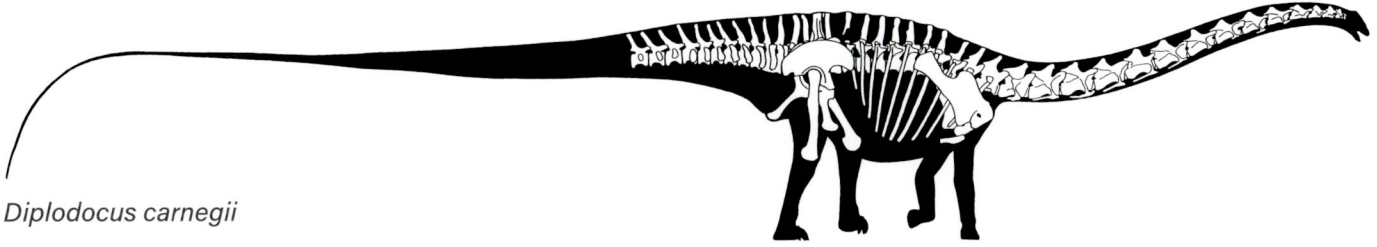

Allosaurus and *Diplodocus*

DISTRIBUTION AND FORMATION/S Wyoming; middle Morrison.
HABITAT Short wet season, otherwise semiarid with open floodplain prairies and riverine forests.
NOTES It is possible this species grew as large as *D. hallorum*.

Tornieria (or *Barosaurus*) *africana*
— 25 m (80 ft) TL, 10 tonnes
FOSSIL REMAINS Minority of skull and several skeletons.
ANATOMICAL CHARACTERISTICS Neck extremely long.
AGE Late Jurassic, late Kimmeridgian and/or early Tithonian.
DISTRIBUTION AND FORMATION/S Tanzania; upper Tendaguru.
HABITAT Coastal, seasonally dry with heavier vegetation inland.
NOTES Remains from much earlier middle Tendaguru are not same taxon.

Barosaurus lentus (see p.132)
— 27 m (88 ft) TL, 13.3 tonnes
FOSSIL REMAINS Possible partial skull, a few partial skeletons.
ANATOMICAL CHARACTERISTICS Neck extremely long. Tail moderately long.
AGE Late Jurassic, early and/or early middle Kimmeridgian.
DISTRIBUTION AND FORMATION/S South Dakota, possibly Wyoming and Utah; middle Morrison.
HABITAT Northern near-coastal portion of range, wetter than rest of Morrison.
HABITS High-level browser, although easily able to graze.
NOTES *Kaatedocus siberi* may belong to this taxon. Presence in more coastal portion of Morrison may be because of presence of taller trees. Partial remains suggest genus reached 30 tonnes.

Barosaurus lentus

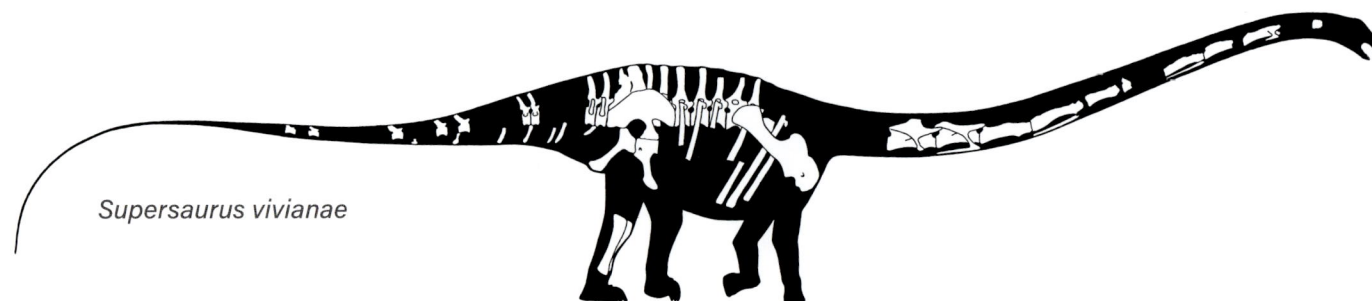

Supersaurus vivianae

Supersaurus vivianae
—— 38 m (125 ft) TL, 50 tonnes
FOSSIL REMAINS Minority of several skeletons.
ANATOMICAL CHARACTERISTICS More robustly built than other diplodocines. Neck very long.
AGE Late Jurassic, early and/or early middle Kimmeridgian.
DISTRIBUTION AND FORMATION/S Colorado; upper middle Morrison.
HABITAT Short wet season, otherwise semiarid with open floodplain prairies and riverine forests.
NOTES Relationships to other diplodocids not entirely certain. Originally incorrectly thought to be the brachiosaur *Ultrasauros* (= *Ultrasaurus*). Estimates of greater bulk problematic.

Supersaurus (= Lourinhasaurus) alenquerensis
—— 18 m (60 ft) TL, 7 tonnes
FOSSIL REMAINS Minority of several skeletons.
ANATOMICAL CHARACTERISTICS Insufficient information.
AGE Late Jurassic, late Kimmeridgian.
DISTRIBUTION AND FORMATION/S Portugal; Camadas de Alcobaça.
HABITAT Large, seasonally dry island with open woodlands.
NOTES May include *Dinheirosaurus lourinhanensis*.

APATOSAURINES
GIGANTIC DIPLODOCIDS LIMITED TO THE LATE JURASSIC OF NORTH AMERICA.

ANATOMICAL CHARACTERISTICS Uniform. Skeletons massively constructed. Necks moderately long. Trunks very short. Tail whips very long. Pelves large to very large.
HABITS Flexible feeder from ground to highest levels. Built for pushing down trees. Powerful build indicates strong defense against predators.

Unnamed genus and species
—— 23 m (75 ft) TL, 13 tonnes
FOSSIL REMAINS Nearly complete skeleton.
ANATOMICAL CHARACTERISTICS Neck fairly long, not very broad, front process of cervical ribs projects forward. Hip sail not especially tall.
AGE Late Jurassic, early late Oxfordian.
DISTRIBUTION AND FORMATION/S Wyoming; lowermost Morrison.
HABITAT Short wet season, otherwise semiarid with open floodplain prairies and riverine forests.

Unnamed genus and species

Elosaurus parvus

Elosaurus parvus
—— 22 m (72 ft) TL, 15 tonnes
FOSSIL REMAINS Majority of adult skeleton, partial juvenile skeleton.
ANATOMICAL CHARACTERISTICS Neck moderately broad, front process of cervical ribs projects forward. Hip sail tall. Pelvis very large.
AGE Late Jurassic, early and/or early middle Kimmeridgian.
DISTRIBUTION AND FORMATION/S Wyoming; middle Morrison.
HABITAT Short wet season, otherwise semiarid with open floodplain prairies and riverine forests.
NOTES Not *Brontosaurus* because lacks its unusual downward projection of front process of cervical ribs, nor later *Apatosaurus* because it lacks its unusually broad neck.

Apatosaurus ajax
—— 23 m (75 ft) TL, 25 tonnes
FOSSIL REMAINS Minority of skeleton.

ANATOMICAL CHARACTERISTICS Neck broad, front process of cervical ribs projects forward.
AGE Late Jurassic, late Kimmeridgian.
DISTRIBUTION AND FORMATION/S Colorado; uppermost Morrison.
HABITAT Wetter than earlier Morrison, otherwise semiarid with open floodplain prairies and riverine forests.
HABITS Broad neck best adapted for horizontal movements.
NOTES Validity of fragmentary original fossil problematic, as is placement of other fossils in this genus.

Unnamed genus and species
—— Adult size uncertain
FOSSIL REMAINS Nearly complete immature skeleton.
ANATOMICAL CHARACTERISTICS Neck apparently very broad, front process of cervical ribs projects forward.

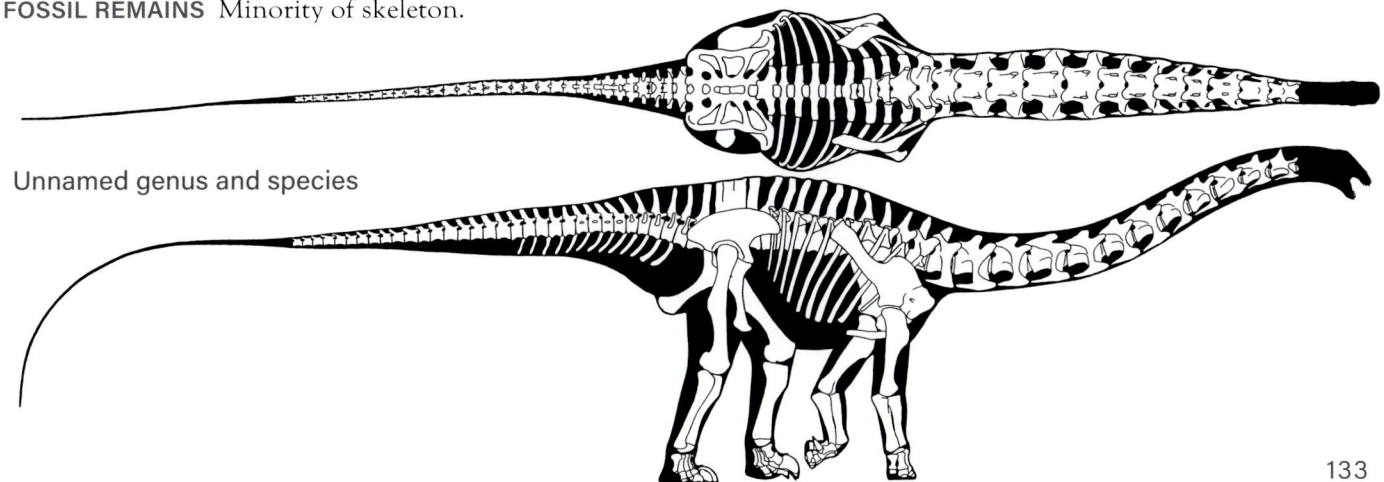

Unnamed genus and species

AGE Late Jurassic, late Kimmeridgian.
DISTRIBUTION AND FORMATION/S Wyoming; upper Morrison.
HABITAT Wetter than earlier Morrison, otherwise semiarid with open floodplain prairies and riverine forests.
HABITS Broad neck best adapted for horizontal movements.
NOTES An apparently broader neck indicates this is not same taxon as later A. *ajax*.

Brontosaurus excelsus
—— 22 m (72 ft) TL, 18 tonnes
FOSSIL REMAINS Majority of skeleton.
ANATOMICAL CHARACTERISTICS Moderately broad neck very deep partly because front process of cervical ribs projects downward. Hip sail tall. Pelvis very large.
AGE Late Jurassic, middle Kimmeridgian.
DISTRIBUTION AND FORMATION/S Wyoming; middle Morrison.

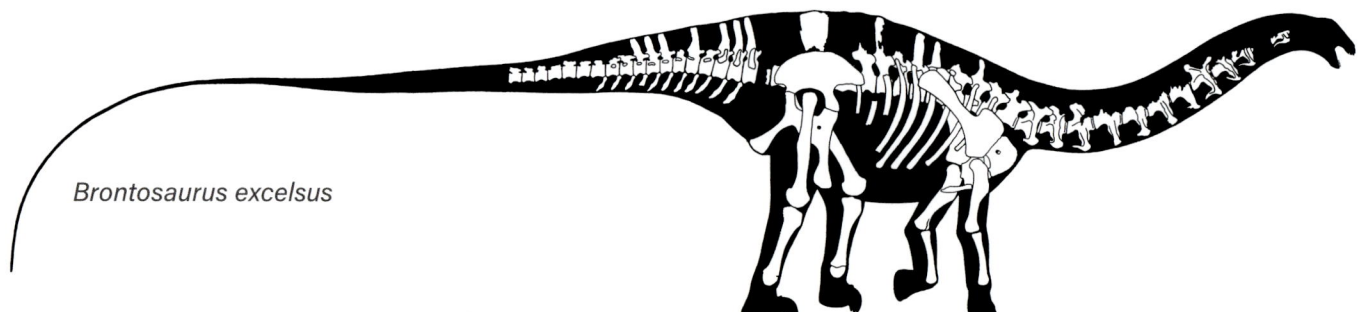

Brontosaurus excelsus

HABITAT Short wet season, otherwise semiarid with open floodplain prairies and riverine forests.
HABITS Deep neck best adapted for vertical movements.
NOTES The classic sauropod genus. Among archosaurs, the front process of cervical ribs projects downward apparently only in *Brontosaurus*.

Brontosaurus louisae?
—— 23 m (75 ft) TL, 20 tonnes
FOSSIL REMAINS Nearly complete skull and skeleton, possibly a few partial skeletons.

ABOVE: *Brontosaurus louisae?*

Brontosaurus louisae?

Brontosaurus louisae?

ANATOMICAL CHARACTERISTICS Moderately broad neck very deep partly because front process of cervical ribs projects downward. Hip sail tall. Pelvis very large.
AGE Late Jurassic, middle Kimmeridgian.
DISTRIBUTION AND FORMATION/S Utah, Wyoming?; later middle Morrison.
HABITAT Short wet season, otherwise semiarid with open floodplain prairies and riverine forests.

HABITS Deep neck best adapted for vertical movements.
NOTES Because the front process of cervical ribs projects downward only in *Brontosaurus*, this is in that genus. Because two skeletons from same quarry are over a million years younger than those of *B. excelsus* and there are minor differences, this may be a different species, but this is not certain. Largest land animal known from a nearly complete individual skeleton.

MACRONARIANS

LARGE TO ENORMOUS NEOSAUROPODS OF THE MIDDLE JURASSIC TO THE END OF THE DINOSAUR ERA, MOST CONTINENTS.

ANATOMICAL CHARACTERISTICS Variable. Nostrils enlarged. Necks able to elevate subvertically. Hands elongated. Pubes broad.
NOTES Absence from Antarctica probably reflects lack of sufficient sampling.

MACRONARIAN MISCELLANEA

NOTES The relationships of these macronarians are uncertain.

Atlasaurus imelakei
— 15 m (50 ft) TL, 23 tonnes
FOSSIL REMAINS Partial skull and majority of skeleton.
ANATOMICAL CHARACTERISTICS Head broad and fairly shallow. Neck rather short. Tail not large. Arm and hand very long, and humerus almost as long as femur, so shoulder much higher than hips. Limbs long relative to size of body.
AGE Middle Jurassic, late Bathonian.

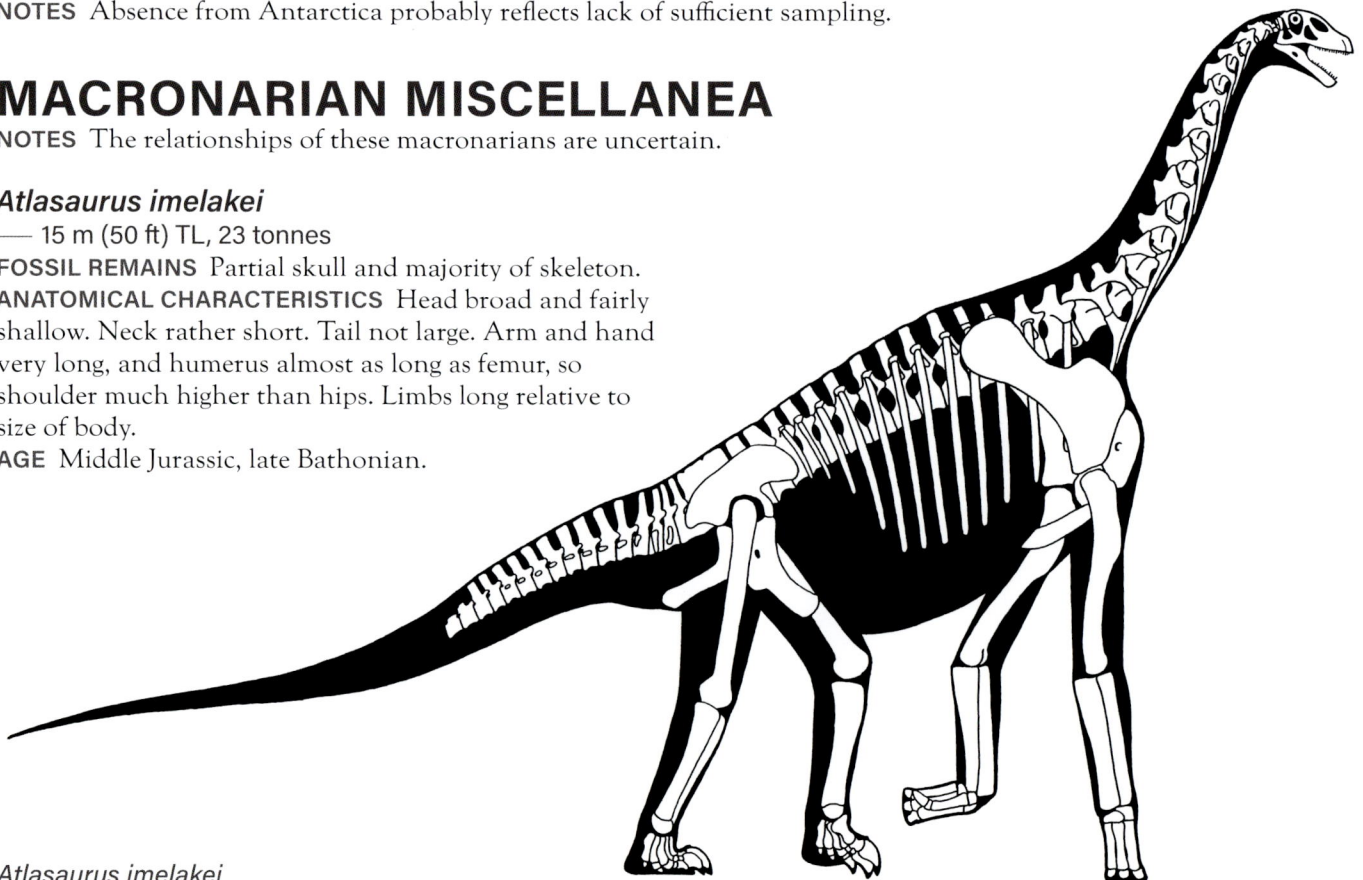

Atlasaurus imelakei

DISTRIBUTION AND FORMATION/S Morocco; Douar of Tazouda.
HABITAT Seasonally arid-wet coastline with tall trees limited to watercourses.
HABITS Medium- and high-level browser, unable to feed easily at ground level.
NOTES Its limbs proportionally longer than those of any other known sauropod, *Atlasaurus* emphasized leg over neck length to increase vertical reach to a greater extent than any other known member of the group. Has been placed in cetiosaurs, turiasaurs, and brachiosaurs.

Abrosaurus dongpoi
—— 11 m (35 ft) TL, 6 tonnes
FOSSIL REMAINS Skull.
ANATOMICAL CHARACTERISTICS Insufficient information.

AGE Middle Jurassic, probably Bathonian.
DISTRIBUTION AND FORMATION/S Central China; lower Xiashaximiao.
HABITAT Heavily forested.

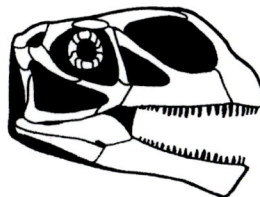

Abrosaurus dongpoi

Abrosaurus dongpoi

Tehuelchesaurus benitezii

—— 15 m (50 ft) TL, 18 tonnes

FOSSIL REMAINS Majority of skeleton, skin patches.
ANATOMICAL CHARACTERISTICS Insufficient
information.
AGE Early Jurassic, middle Toarcian.
DISTRIBUTION AND FORMATION/S Southern Argentina;
upper Cañadón Asfalto.
HABITAT Short wet season, otherwise semiarid, riverine
forests, open floodplains.

Jobaria tiguidensis

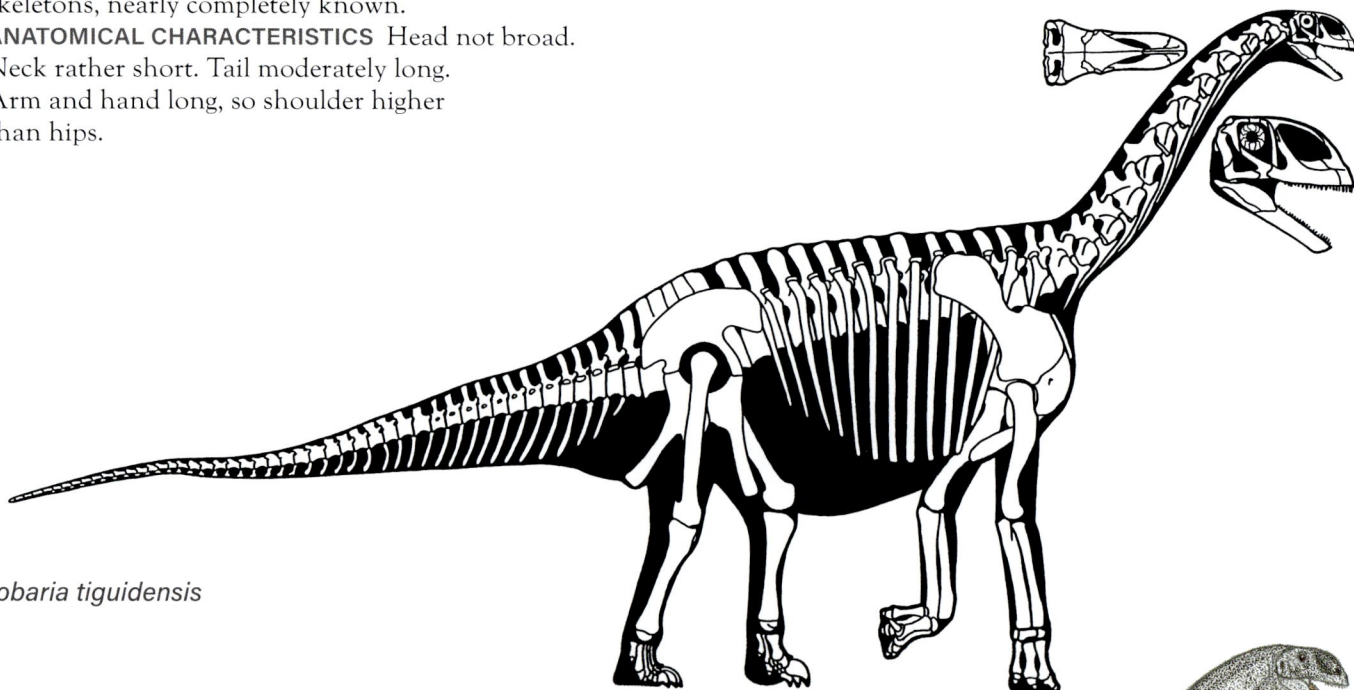

—— 16 m (52 ft) TL, 18 tonnes

FOSSIL REMAINS Complete skull and several
skeletons, nearly completely known.
ANATOMICAL CHARACTERISTICS Head not broad.
Neck rather short. Tail moderately long.
Arm and hand long, so shoulder higher
than hips.

AGE Late Middle or early
Late Jurassic.
DISTRIBUTION AND FORMATION/S Niger; Tiouraren.
HABITAT Well-watered woodlands.
HABITS Medium- and high-level browser, unable to feed
easily at ground level.
NOTES Originally thought to be from the Early
Cretaceous, the Tiouraren is from the later Jurassic.
Relationships of *Jobaria* are uncertain, may not be a
neosauropod.

Jobaria tiguidensis

Jobaria tiguidensis

Yuzhoulong qurenensis
—— Adult size uncertain
FOSSIL REMAINS Partial skull and skeleton, immature.
ANATOMICAL CHARACTERISTICS Insufficient information.
AGE Late Jurassic, probably Bathonian.
DISTRIBUTION AND FORMATION/S Central China; lower Xiashaximiao.
HABITAT Heavily forested.

Janenschia robusta
—— 17 m (53 ft) TL, 20 tonnes
FOSSIL REMAINS Minority of a few skeletons.
ANATOMICAL CHARACTERISTICS Fingers and thumb claw present.
AGE Late Jurassic, late Kimmeridgian and/or early Tithonian.
DISTRIBUTION AND FORMATION/S Tanzania; upper Tendaguru.

HABITAT Coastal, seasonally dry with heavier vegetation inland.
NOTES Remains from much earlier middle Tendaguru are not same taxon. Previously considered the only named titanosaur from the Jurassic.

Haestasaurus becklesii
—— Adult size uncertain
FOSSIL REMAINS Minority of skeleton, immature, skin patch.
ANATOMICAL CHARACTERISTICS Forelimb robust.
AGE Early Cretaceous, Berriasian or Valanginian.
DISTRIBUTION AND FORMATION/S Southeastern England; Hastings Beds.
NOTES May be a titanosaur. Includes elements once assigned to *Pelorosaurus brevis*. One of the few examples of sauropod skin.

Jobaria tiguidensis

CAMARASAURIDS

LARGE TO GIGANTIC MACRONARIAN SAUROPODS LIMITED TO THE LATE JURASSIC TO PERHAPS THE EARLY CRETACEOUS OF NORTH AMERICA AND EUROPE.

ANATOMICAL CHARACTERISTICS Uniform. Heads large for sauropods, deep, teeth fairly large. Necks rather short, shallow, broad. Most neck and trunk vertebral spines forked. Tails moderately long. Arms and hands long, so shoulders a little higher than hips. Front of pelvis and belly ribs flare very strongly sideways, so belly is very broad and large. Retroverted pelves facilitated slow walking when rearing up by keeping hips and tails horizontal when bipedal.

HABITS Medium- and high-level browsers, unable to feed easily at ground level. Able to consume coarse vegetation. Little change in head proportions with growth.

NOTES Whether camarasaurs survived into the Early Cretaceous is uncertain.

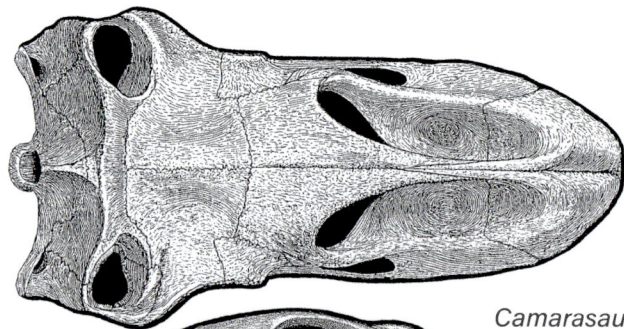

Camarasaurus shaded skull

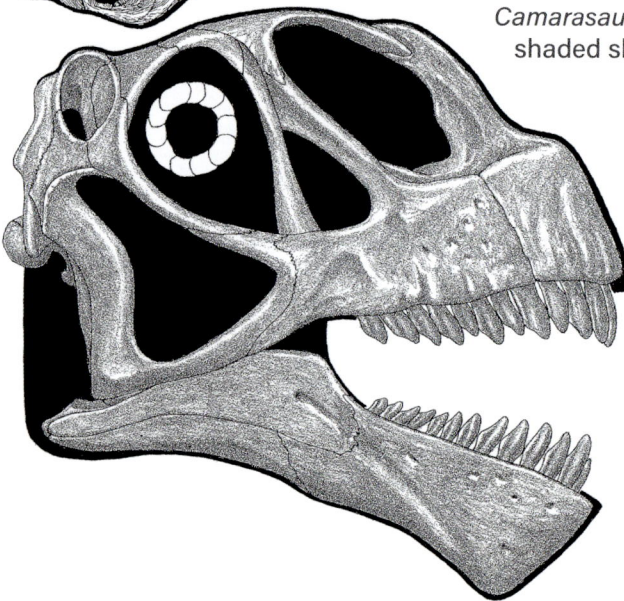

Camarasaurus grandis

Lack of a complete adult and uncertain assignment of some skeletons preclude skeletal restoration.

Camarasaurus lentus
—— 15 m (50 ft) TL, 16 tonnes
FOSSIL REMAINS A number of skulls and skeletons, including juveniles, completely known.
ANATOMICAL CHARACTERISTICS Standard for group.
AGE Late Jurassic, middle Kimmeridgian.
DISTRIBUTION AND FORMATION/S Wyoming, Colorado, Utah; upper middle Morrison.
HABITAT Short wet season, otherwise semiarid with open floodplain prairies and riverine forests.
NOTES May be the direct ancestor of *C. supremus*.

Camarasaurus supremus (see p.143)
—— 18 m (60 ft) TL, 24 tonnes
FOSSIL REMAINS Some skulls and skeletons.
ANATOMICAL CHARACTERISTICS Standard for group.
AGE Late Jurassic, late middle and/or late Kimmeridgian.
DISTRIBUTION AND FORMATION/S Wyoming, Colorado, New Mexico; upper Morrison.
HABITAT Wetter than earlier Morrison, otherwise semiarid with open floodplain prairies and riverine forests.

Camarasaurus grandis
—— 14 m (45 ft) TL, 13 tonnes
FOSSIL REMAINS A few skulls and majority of skeletons, some juveniles.
ANATOMICAL CHARACTERISTICS Standard for group.
AGE Late Jurassic, early Kimmeridgian.
DISTRIBUTION AND FORMATION/S Wyoming, Colorado, Montana; middle Morrison.
HABITAT Short wet season, otherwise semiarid with open floodplain prairies and riverine forests.
NOTES Apparently present earlier in the Middle Morrison than *C. lentus* and may be its direct ancestor.

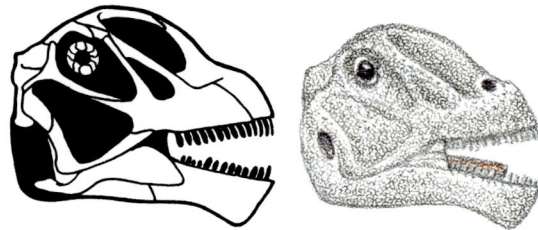

Camarasaurus (= Cathetosaurus) lewisi (see p.143)
—— 13 m (43 ft) TL, 12 tonnes
FOSSIL REMAINS A complete skull and skeleton, majority of a skeleton,.

ANATOMICAL CHARACTERISTICS Modestly sized. Head quite large, short, deep. Trunk short, belly deep. Pelvis deep, more strongly retroverted.
AGE Late Jurassic, early and/or early middle Kimmeridgian.
DISTRIBUTION AND FORMATION/S Colorado, Wyoming; middle Morrison.
HABITAT Short wet season, otherwise semiarid with open floodplain prairies and riverine forests.
NOTES Relatively largest head among sauropods. Best adapted among known sauropods for walking bipedally.

Aragosaurus ischiaticus
—— 18 m (60 ft) TL, 25 tonnes
FOSSIL REMAINS Minority of skeleton.
ANATOMICAL CHARACTERISTICS Arm longer than that of *Camarasaurus*, so shoulder higher.
AGE Early Cretaceous, late Hauterivian and/or early Barremian.
DISTRIBUTION AND FORMATION/S Northern Spain; El Castellar.
HABITS High-level browser.
NOTES Relationships uncertain.

Camarasaurus lentus
(see also next page)

Camarasaurus lentus

Camarasaurus supremus

Camarasaurus lewisi

TITANOSAURIFORMS

LARGE TO ENORMOUS MACRONARIAN SAUROPODS OF THE MIDDLE OR LATE JURASSIC TO THE END OF THE DINOSAUR ERA, MOST CONTINENTS.

ANATOMICAL CHARACTERISTICS Variable. Teeth elongated. Gauge of trackways broader than those of other sauropods. Front of pelves and belly ribs flare very strongly sideways, so bellies are very broad and large. Fingers further reduced or absent, thumb claws reduced or absent.

NOTES Absence from Antarctica probably reflects lack of sufficient sampling.

TITANOSAURIFORM MISCELLANEA

NOTES: THE RELATIONSHIPS OF THESE TITANOSAURIFORMS ARE UNCERTAIN.

Unnamed genus and species

—— 20 m (65 ft) TL, 30 tonnes

FOSSIL REMAINS Majority of skeleton.

ANATOMICAL CHARACTERISTICS Humerus shorter than femur, hand moderately long.

AGE Late Jurassic, early and/or early middle Kimmeridgian.

DISTRIBUTION AND FORMATION/S Wyoming; lower middle Morrison.

HABITAT Short wet season, otherwise semiarid with open floodplain prairies and riverine forests.

NOTES From Como Bluff.

Oceanotitan dantasi

—— Adult size uncertain

FOSSIL REMAINS Minority of skeleton, probably immature.

ANATOMICAL CHARACTERISTICS Insufficient information.

AGE Late Jurassic, upper Kimmeridgian and/or lowermost Tithonian.

DISTRIBUTION AND FORMATION/S Portugal; lower Lourinha.

HABITAT Large, seasonally dry island with open woodlands.

NOTES May not be a titanosauriform.

Duriatitan humerocristatus

—— Adult size uncertain

FOSSIL REMAINS Minority of skeleton, possibly immature.

ANATOMICAL CHARACTERISTICS Insufficient information.

AGE Late Jurassic, early Kimmeridgian.

DISTRIBUTION AND FORMATIONS Southern England; lower Kimmeridge Clay.

HABITAT Island archipelago shallows.

NOTES Found as drift in coastal sediments. May be an island dwarf.

Australodocus bohetii

—— 17 m (55 ft) TL, 4 tonnes

FOSSIL REMAINS Neck vertebrae.

ANATOMICAL CHARACTERISTICS Neck very long.

AGE Late Jurassic, late Kimmeridgian and/or early Tithonian.

DISTRIBUTION AND FORMATION/S Tanzania; upper Tendaguru.

HABITAT Coastal, seasonally dry with heavier vegetation inland.

NOTES Long considered a diplodocid.

Fusuisaurus zhaoi

—— 22 m (70 ft) TL, 35 tonnes

FOSSIL REMAINS Minority of skeleton.

ANATOMICAL CHARACTERISTICS Insufficient information.

AGE Early Cretaceous.

DISTRIBUTION AND FORMATION/S Southern China; Napai.

Huanghetitan liujiaxiaensis

—— 12 m (40 ft) TL, 3 tonnes

FOSSIL REMAINS Minority of skeleton.

ANATOMICAL CHARACTERISTICS Insufficient information.

AGE Early Cretaceous, Aptian or Albian.

DISTRIBUTION AND FORMATION/S Northern China; Haoling.

Ruixinia zhangi

—— 12 m (40 ft) TL, 3 tonnes

FOSSIL REMAINS Partial skeleton.

ANATOMICAL CHARACTERISTICS Neck moderately long. Tail moderately long, last half dozen vertebrae fused into slender rod.

AGE Early Cretaceous, latest Barremian.

DISTRIBUTION AND FORMATION/S Northeastern China; lower Yixian.

HABITAT Well-watered, volcanic, highland forest-and-lake district, winters chilly with snow.

HABITS Function of tail rod uncertain.

Dongbeititan dongi
—— 15 m (59 ft) TL, 7 tonnes
FOSSIL REMAINS Minority of skeleton.
ANATOMICAL CHARACTERISTICS Heavily constructed. Neck broad, moderately long.
AGE Early Cretaceous, latest Barremian and/or earliest Aptian.
DISTRIBUTION AND FORMATION/S Northeastern China; middle Yixian.
HABITAT Well-watered, volcanic, highland forest-and-lake district, winters chilly with snow.

Yongjinglong datangi
—— Adult size uncertain
FOSSIL REMAINS Minority of skeleton/s.
ANATOMICAL CHARACTERISTICS Insufficient information.
AGE Late Early Cretaceous.
DISTRIBUTION AND FORMATION/S Northern China; Hekou Group?
NOTES Size disparities between elements indicate elements are not from one skeleton. Is not certain whether this and nearby *Daxiatitan* and *Huanghetitan* are from the same formation.

Garumbatitan morellensis
—— 20 m (65 ft) TL, 30 tonnes
FOSSIL REMAINS Two partial skeletons, fragmentary remains.
ANATOMICAL CHARACTERISTICS Insufficient information.
AGE Early Cretaceous, late Barremian.
DISTRIBUTION AND FORMATION/S Eastern Spain; upper Arcillas de Morella.

Tastavinsaurus sanzi
—— 16 m (50 ft) TL, 8 tonnes
FOSSIL REMAINS Minority of skeleton.
ANATOMICAL CHARACTERISTICS Insufficient information.
AGE Early Cretaceous, early Aptian.
DISTRIBUTION AND FORMATION/S Eastern Spain; Xert.

Fukuititan nipponensis
—— Adult size uncertain
FOSSIL REMAINS Minority of skeleton.
ANATOMICAL CHARACTERISTICS Insufficient information.
AGE Early Cretaceous, Barremian.
DISTRIBUTION AND FORMATION/S Central Japan; Kitadani.

Wintonotitan wattsi
—— 15 m (50 ft) TL, 10 tonnes
FOSSIL REMAINS Minority of two skeletons.
ANATOMICAL CHARACTERISTICS Insufficient information.
AGE Late Cretaceous, late Cenomanian and/or early Turonian.
DISTRIBUTION AND FORMATION/S Northeastern Australia; upper Winton.
HABITAT Well-watered areas, cold winters.

EUHELOPIDS

SMALL (FOR SAUROPODS) TO GIGANTIC TITANOSAURIFORM SAUROPODS OF THE EARLY OR LATE CRETACEOUS OF ASIA.

ANATOMICAL CHARACTERISTICS Fairly uniform. Heads fairly broad, snouts form a shelf below nostrils, which are very large and arced. Skeletons rather lightly built. Necks moderately to very long. Tails not large. Arms and hands very to exceptionally long, so shoulder much higher than hips, thumb claws usually reduced or absent. Pelves rather small, retroverted.
HABITS High-level browsers, unable to feed easily near ground level.
NOTES Contents of this family not certain.

Euhelopus zdanskyi (see pp.146–47)
—— 11 m (35 ft) TL, 3.8 tonnes
FOSSIL REMAINS Majority of skull and two skeletons.
ANATOMICAL CHARACTERISTICS Neck long. Vertebral spines near base of neck forked. Arm probably very long, so shoulder higher than hips.
AGE Early Cretaceous, Berriasian or Valanginian.
DISTRIBUTION AND FORMATION/S Eastern China; Mengyin.
NOTES Proportions of arm and leg uncertain. Long thought a Late Jurassic mamenchisaur relative, now often an Early Cretaceous titanosauriform.

Euhelopus zdanskyi

Silutitan sinensis

—— 15 m (50 ft) TL, 5 tonnes

FOSSIL REMAINS Minority of one or two skeletons.
ANATOMICAL CHARACTERISTICS Neck moderately long.
AGE Early Cretaceous.
DISTRIBUTION AND FORMATION/S Northeastern China; Shengjinkou.
HABITAT Well-watered forests and lakes, winters chilly with some snow.
NOTE/S May include *Hamititan xinjiangensis*.

Erketu ellisoni

—— 15 m (50 ft) TL, 5 tonnes

FOSSIL REMAINS Minority of skeleton.
ANATOMICAL CHARACTERISTICS Neck extremely long, with vertebrae more elongated than in any other sauropod.
AGE Early Cretaceous.
DISTRIBUTION AND FORMATION/S Mongolia; Bayanshiree.
HABITS A high browser.

Euhelopus zdanskyi

Phuwiangosaurus sirindhornae

—— 19 m (62 ft) TL, 17 tonnes

FOSSIL REMAINS Partial skeletons, juvenile and adult.
ANATOMICAL CHARACTERISTICS Neck moderately long, some vertebral spines forked.
AGE Early Cretaceous, Valanginian or Hauterivian.
DISTRIBUTION AND FORMATION/S Thailand; Sao Khua.
HABITAT Short wet season, otherwise semiarid with open floodplain prairies and riverine forests.

Tangvayosaurus hoffeti

—— 19 m (62 ft) TL, 17 tonnes

FOSSIL REMAINS Two partial skeletons.
ANATOMICAL CHARACTERISTICS Skeleton robustly built.
AGE Early Cretaceous, late Aptian or Albian.
DISTRIBUTION AND FORMATION/S Laos; Grès Supérieurs.

Qiaowanlong kangxii

—— 12 m (40 ft) TL, 6 tonnes

FOSSIL REMAINS Minority of skeleton.
ANATOMICAL CHARACTERISTICS Neck long, vertebral spines forked.
AGE Early Cretaceous, Aptian or Albian.
DISTRIBUTION AND FORMATION/S Central China; middle Xinminpu.

Yunmenglong ruyangensis

—— 20 m (65 ft) TL, 30 tonnes

FOSSIL REMAINS Minority of skeleton.
ANATOMICAL CHARACTERISTICS Neck very long.
AGE Early Cretaceous, Aptian or Albian.
DISTRIBUTION AND FORMATION/S Eastern China; Haoling.

Tambatitanis amicitiae

—— 14 m (45 ft) TL, 4 tonnes

FOSSIL REMAINS Minority of skeleton.
ANATOMICAL CHARACTERISTICS Insufficient information.
AGE Early Cretaceous, probably early Albian.
DISTRIBUTION AND FORMATION/S Central Japan; lower Sasayama Group.

Daxiatitan bingling

—— Adult size not available

FOSSIL REMAINS Minority of skeleton.
ANATOMICAL CHARACTERISTICS Insufficient information.
AGE Late Early Cretaceous.
DISTRIBUTION AND FORMATION/S Northern China; Hekou Group.

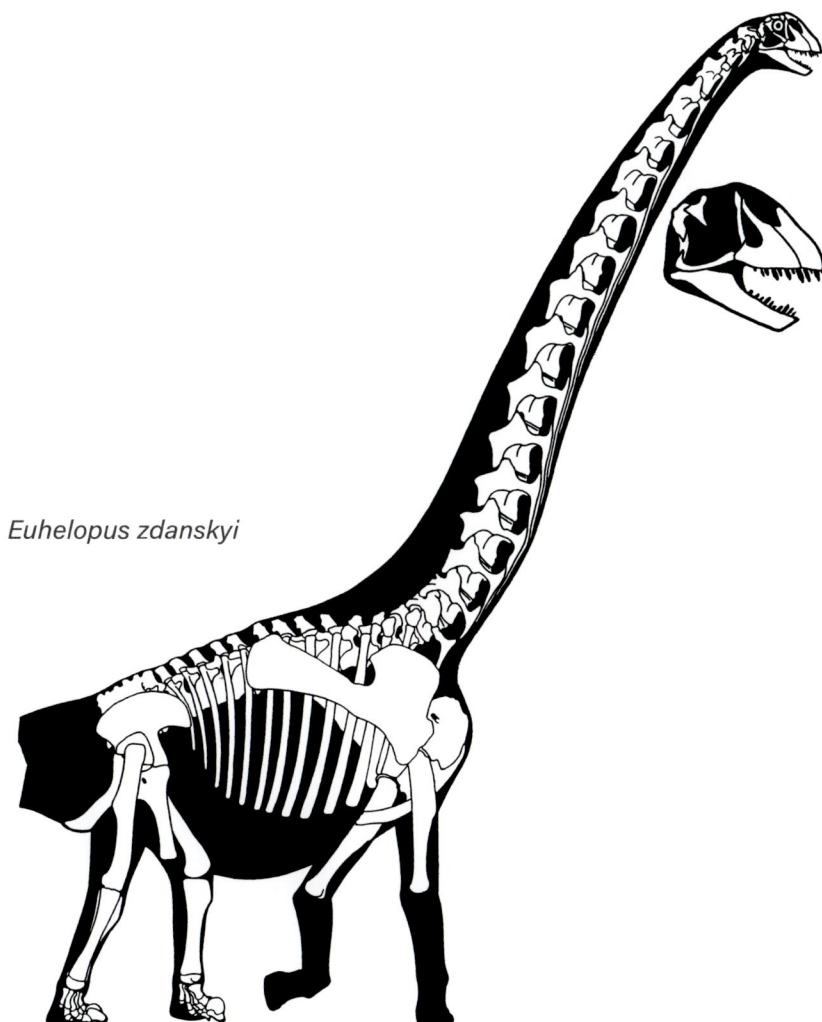

Jiangshanosaurus lixianensis
—— 11 m (35 ft) TL, 2.5 tonnes
FOSSIL REMAINS Minority of skeleton.
ANATOMICAL CHARACTERISTICS Insufficient information.
AGE Early Cretaceous, Turonian or Coniacian.
DISTRIBUTION AND FORMATION/S Southeastern China; Jinhua.

Ligabuesaurus leanzai
—— 18 m (60 ft) TL, 20 tonnes
FOSSIL REMAINS Minority of skull and two partial skeletons.
ANATOMICAL CHARACTERISTICS Neck moderately long. Spines of neck and trunk vertebrae very broad. Shoulder girdle appears to be relatively small, so shoulders may not have been high-set despite long arm.
AGE Early Cretaceous, late Aptian or early Albian.
DISTRIBUTION AND FORMATION/S Western Argentina; Lohan Cura.
HABITAT Well-watered coastal woodlands with short dry season.

Huabeisaurus allocotus
—— 18 m (60 ft) TL, 11 tonnes
FOSSIL REMAINS Majority of skeleton.
ANATOMICAL CHARACTERISTICS Neck long. Shoulders level with hips.
AGE Late Cretaceous.
DISTRIBUTION AND FORMATION/S Northern China; upper Huiquanpu.
NOTES Relationships uncertain, if a euhelopid, indicates presence of group in Late Cretaceous.

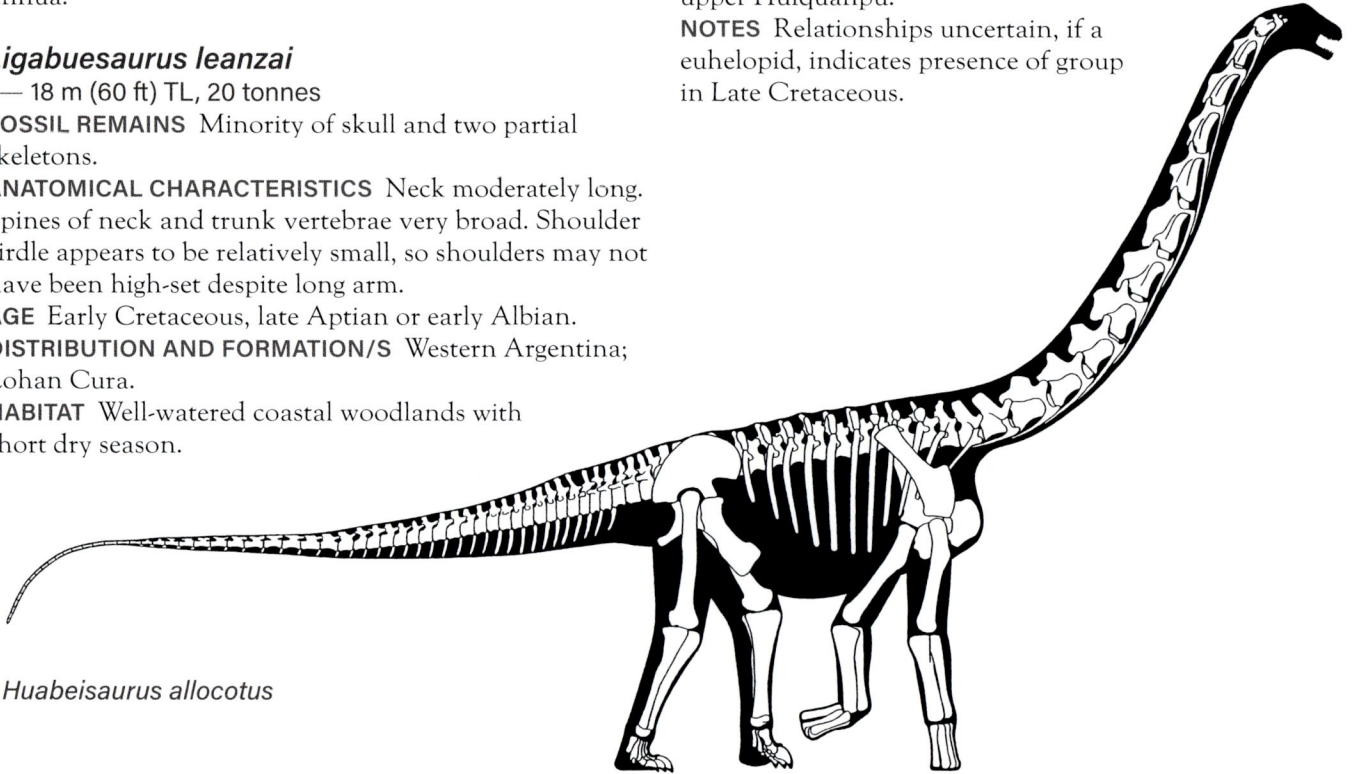

Huabeisaurus allocotus

Ruyangosaurus giganteus
—— 25 (80 ft) TL, 35 tonnes
FOSSIL REMAINS Minority of skeleton.
ANATOMICAL CHARACTERISTICS Insufficient information.

AGE Early Cretaceous, Aptian or Albian.
DISTRIBUTION AND FORMATION/S Eastern China; Haoling.
NOTES Relationships uncertain.

BRACHIOSAURIDS

SMALL (FOR SAUROPODS) TO ENORMOUS TITANOSAURIFORM SAUROPODS LIMITED TO THE LATE JURASSIC AND EARLY CRETACEOUS OF THE AMERICAS, EUROPE, AND AFRICA.

ANATOMICAL CHARACTERISTICS Fairly uniform. Heads fairly broad, snouts form shelf below nostrils, which are very large and arced. Skeletons rather lightly built. Necks moderately to very long. Tails not large. Arms and hands very to exceptionally long, so shoulder much higher than hips. Thumb claws reduced or absent. Pelves rather small, retroverted.
HABITS High-level browsers, unable to feed easily near ground level. Reared up less often than other sauropods. Fragmentary Early Cretaceous Colombian *Padillasaurus leivaensis* indicates presence of group in South America.

Giraffatitan shaded skull

Giraffatitan muscle study

Vouivria damparisensis
—— 15 m (50 ft) TL, 10 tonnes
FOSSIL REMAINS Partial skeleton.
ANATOMICAL CHARACTERISTICS Insufficient information.
AGE Late Jurassic, late middle and/or early late Oxfordian.
DISTRIBUTION AND FORMATION/S Eastern France; Calcaires de Clerval.

Europasaurus holgeri (see p.150)
—— 5.7 m (19 ft) TL, 830 kg (1,800 lb)
FOSSIL REMAINS Majority of skull and a number of skeletons.
ANATOMICAL CHARACTERISTICS Snout shelf short. Neck moderately long. Thumb claw small.
AGE Late Jurassic, middle Kimmeridgian.

DISTRIBUTION AND FORMATION/S Northern Germany; Mittlere Kimmeridge-Stufe.
HABITS Small size limited browsing height.
NOTES Found as drift in nearshore marine sediments set amid islands. Small size is probably dwarfism forced by limited food resources. May not be a brachiosaurid.

Europasaurus holgeri

149

Europasaurus holgeri

Brachiosaurus altithorax
—— 22 m (72 ft) TL, 38 tonnes
FOSSIL REMAINS Minority of skeleton, possible partial skull, other bones.
ANATOMICAL CHARACTERISTICS Tail short for sauropods. Arm and hand exceptionally long, and humerus longer than femur, so shoulders very high.
AGE Late Jurassic, middle Kimmeridgian.
DISTRIBUTION AND FORMATION/S Colorado; middle Morrison.
HABITAT Short wet season, otherwise semiarid with open floodplain prairies and riverine forests.
NOTES Probably includes *Dystylosaurus edwini*. Some Morrison brachiosaur remains are more similar to *Giraffatitan*.

Giraffatitan brancai
—— 25 m (80 ft) TL, 50 tonnes
FOSSIL REMAINS At least one partial skull and skeleton, possibly other skulls and skeletons.

ANATOMICAL CHARACTERISTICS Snout shelf long, head a little hourglass-shaped in top view. Neck very long. Tall withers at shoulder anchored unusually deep neck ligaments and tendons. Back trunk vertebrae relatively small. Tail short for sauropods. Arm and hand exceptionally long, and humerus longer than femur, so shoulders very high, limbs long relative to body. Thumb claw small.
AGE Late Jurassic, middle Kimmeridgian.
DISTRIBUTION AND FORMATION/S Tanzania; middle Tendaguru.
HABITAT Coastal, seasonally dry with heavier vegetation farther inland.
NOTES The most giraffe-like dinosaur known, both neck and limb length used to increase vertical reach. Not placeable in the *Brachiosaurus* to which it was long assigned. A portion of remains placed in *G. brancai* from middle Tendaguru may be different taxa, those from much earlier lower Tendaguru and much later upper Tendaguru more so. One fossil preserved in position suggests it died stranded in tidal mudflat.

Giraffatitan brancai

Brachiosaurus

Lusotitan atalaiensis
—— 21 m (70 ft) TL, 25 tonnes
FOSSIL REMAINS Minority of skeletons.
ANATOMICAL CHARACTERISTICS Humerus longer than femur, so shoulders very high.
AGE Late Jurassic, early or middle Tithonian.
DISTRIBUTION AND FORMATION/S Portugal; middle Lourinhã.
HABITAT Large, seasonally dry island with open woodlands.
NOTES Relationships of *Lusotitan* uncertain. The presence of this and other gigantic sauropods on a Portuguese island shows that dwarfism was not occurring, perhaps because of intermittent immigration from nearby continents.

Galvesaurus herreroi
—— Adult size uncertain
FOSSIL REMAINS Minority of several skeletons.
ANATOMICAL CHARACTERISTICS Neck long.
AGE Late Jurassic, late Kimmeridgian and/or early Tithonian.

DISTRIBUTION AND FORMATION/S Eastern Spain; lower Villar del Arzobispo.
NOTES Subadult remains indicate a very large sauropod.

Cedarosaurus weiskopfae
—— 15 m (50 ft) TL, 10 tonnes
FOSSIL REMAINS Majority of skeleton.
ANATOMICAL CHARACTERISTICS Neck length uncertain.
AGE Early Cretaceous, upper Valanginian.
DISTRIBUTION AND FORMATION/S Utah; lower Cedar Mountain.
HABITAT Short wet season, otherwise semiarid with floodplain prairies, open woodlands, and riverine forests.

Abydosaurus mcintoshi
—— Adult size uncertain
FOSSIL REMAINS Complete skull and partial skull and skeletal remains.
ANATOMICAL CHARACTERISTICS Snout shelf long, nasal opening and projection moderately developed.
AGE Early Cretaceous, upper Albian.

DISTRIBUTION AND FORMATION/S Utah; middle Cedar Mountain.
HABITAT Short wet season, otherwise semiarid with floodplain prairies, open woodlands, and riverine forests.

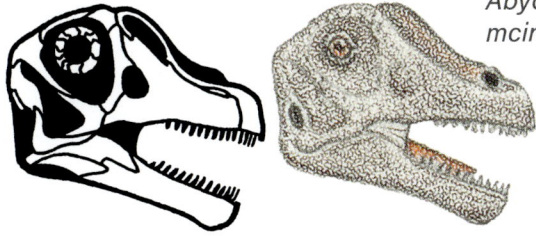

Abydosaurus mcintoshi

Soriatitan golmayensis
—— Adult size uncertain
FOSSIL REMAINS Minority of skeleton, possibly immature.
ANATOMICAL CHARACTERISTICS Standard for group.
AGE Early Cretaceous, late Hauterivian or early Barremian.
DISTRIBUTION AND FORMATION/S Northeastern Spain; Golmayo.

Sonorasaurus thompsoni
—— 15 m (50 ft) TL, 10 tonnes
FOSSIL REMAINS Small minority of skeleton(s).
ANATOMICAL CHARACTERISTICS Toe claws reduced.
AGE Early Cretaceous, late Albian.

DISTRIBUTION AND FORMATION/S Arizona; middle Turney Ranch.

Venenosaurus dicrocei
—— 12 m (40 ft) TL, 5 tonnes
FOSSIL REMAINS Minority of skeleton.
ANATOMICAL CHARACTERISTICS Insufficient information.
AGE Early Cretaceous, late Barremian.
DISTRIBUTION AND FORMATION/S Utah; middle Cedar Mountain.
HABITAT Short wet season, otherwise semiarid with floodplain prairies, open woodlands, and riverine forests.

Pleurocoelus nanus
Adult size uncertain
FOSSIL REMAINS Minority of a few juvenile skulls and skeletons.

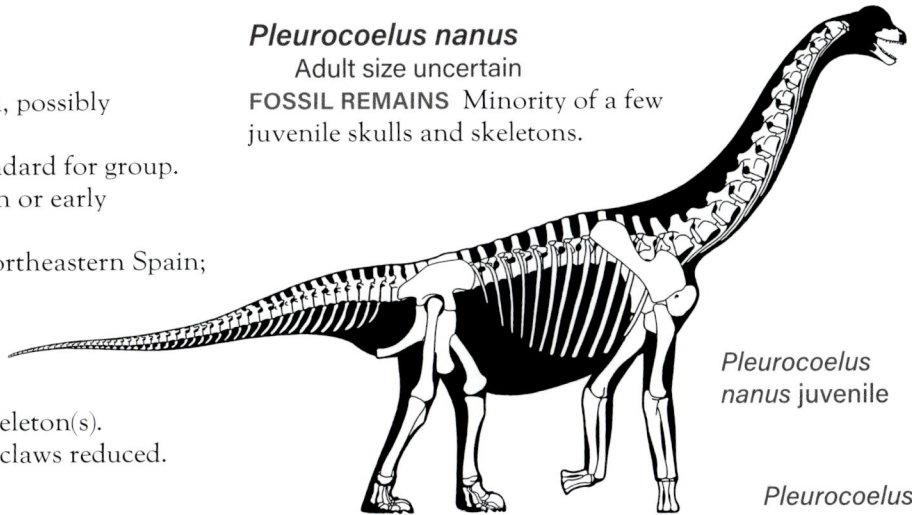

Pleurocoelus nanus juvenile

Pleurocoelus nanus

Sauroposeidon and *Acrocanthosaurus*

ANATOMICAL CHARACTERISTICS Neck moderately long in juveniles.
AGE Early Cretaceous, middle or late Aptian or early Albian.
DISTRIBUTION AND FORMATION/S Maryland; Arundel.
NOTES Originally *Astrodon johnstoni* based on inadequate remains.

Sauroposeidon proteles
—— 27 m (90 ft) TL, 45 tonnes
FOSSIL REMAINS Several partial skeletons.
ANATOMICAL CHARACTERISTICS Neck long.
AGE Early Cretaceous, Aptian.
DISTRIBUTION AND FORMATION/S Texas, Oklahoma; Antlers, Paluxy, Glen Rose.

HABITAT Floodplain with coastal swamps and marshes.
NOTES Includes *Paluxysaurus jonesi*. May not be a brachiosaurid.

Europatitan eastwoodi
—— 20 m (65 ft) TL, 25 tonnes
FOSSIL REMAINS Minority skeletons.
ANATOMICAL CHARACTERISTICS Insufficient information.
AGE Early Cretaceous, late Barremian or early Aptian.
DISTRIBUTION AND FORMATION/S Spain; Castrillo de la Reina.
NOTES Inclusion of two enormous cervicals found near skeleton is incorrect because they are much too large for it.

LATE TITANOSAURIFORM MISCELLANEA

Liaoningotitan sinensis
—— Adult size uncertain
FOSSIL REMAINS Partial skull and skeleton.
ANATOMICAL CHARACTERISTICS Lower jaw strongly downturned.
AGE Early Cretaceous, late Barremian or early Aptian.

DISTRIBUTION AND FORMATION/S Northeastern China, Yixian.
HABITAT Well-watered, volcanic, highland forest-and-lake district, winters chilly with snow.
NOTES May or may not be a titanosaur.

154

TITANOSAURIDS

LARGE TO ENORMOUS TITANOSAURIFORMS OF THE CRETACEOUS UP TO THE END OF THE DINOSAUR ERA, MOST CONTINENTS.

ANATOMICAL CHARACTERISTICS Fairly uniform. Heads long, shallow, bony nostrils strongly retracted to above the orbits, but fleshy nostrils probably still near front of snout. Lower jaws short, shallow, pencil-shaped teeth limited to front of jaws. Heads flexed downward relative to necks. Trunk vertebrae more flexible, possibly aiding rearing. Tails moderately long, very flexible especially upward, ending in a short whip. Arms at least fairly long, so shoulders as high as or higher than hips. Often armored, usually lightly, especially in adults.

HABITS Often used armor as the passive side of their defense strategy, may have been most important in the more vulnerable juveniles. Flexible tail may have been used as display organ by arcing it over the back. Fossil dung indicates titanosaurs consumed flowering plants, including early grasses, as well as nonflowering plants.

NOTES Absence from Antarctica probably reflects lack of sufficient sampling. The last of the sauropod groups, titanosaurs are the only sauropods known to have survived into the late Late Cretaceous. Armor may have assisted them in surviving in a world of increasingly sophisticated and gigantic predators. The relationships of the numerous but often incompletely preserved titanosaurs are not well understood, the group is potentially splittable into a number of subdivisions. A 2m long (6.5 ft) probable shank bone, since disintegrated, from the latest Cretaceous Kallamedu Formation in India labeled *Bruhathkayosaurus matleyi* suggests a titanosaur of 120–200 tonnes, in which case it may be the largest known animal species and indicates that supersized sauropods survived to the close of the dinosaur era.

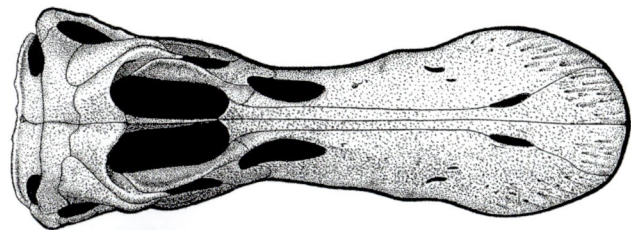

Nemegtosaurus
shaded skull

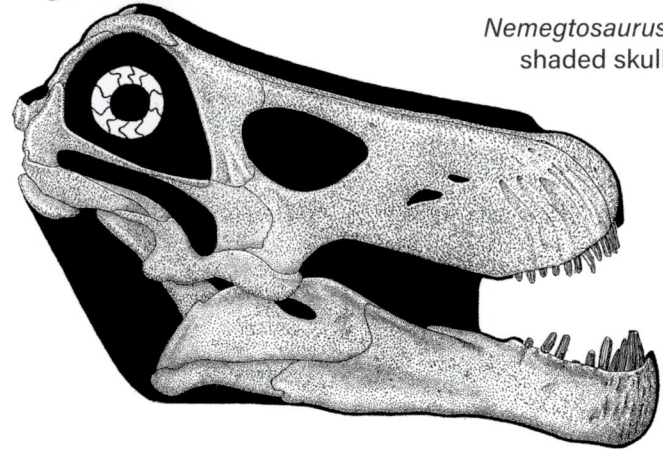

BASO-TITANOSAURIDS

LARGE TO ENORMOUS TITANOSAURIDS OF THE CRETACEOUS UP TO THE END OF THE DINOSAUR ERA, MOST CONTINENTS.

Mnyamawamtuka moyowamkia
—— Adult size uncertain
FOSSIL REMAINS Partial skeleton, possibly immature.
ANATOMICAL CHARACTERISTICS Neck moderately long.
AGE Late Early Cretaceous or early Late Cretaceous.
DISTRIBUTION AND FORMATION/S Tanzania; lower Galula.

Rukwatitan bisepultus
—— Adult size uncertain
FOSSIL REMAINS Partial skeleton, possibly immature.
ANATOMICAL CHARACTERISTICS Neck long.
AGE Late Cretaceous.
DISTRIBUTION AND FORMATION/S Tanzania; upper Galula.
NOTES *Shingopana songwensis* may be juvenile of this taxon.

Sarmientosaurus musacchioi
—— Adult size uncertain
FOSSIL REMAINS Skull and minority of skeleton, probably immature.

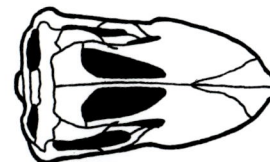

Sarmientosaurus musacchioi

155

ANATOMICAL CHARACTERISTICS Head broad in top view, snout somewhat pointed. Lower jaw not short, teeth in front half of upper jaw and fairly large.
AGE Late Cretaceous, late Cenomanian or Turonian.
DISTRIBUTION AND FORMATION/S Southern Argentina; lower Bajo Barreal.
HABITAT Seasonally wet, well-forested floodplain.
NOTES Head represents a basal titanosaur condition.

Dongyangosaurus sinensis
—— 15 m (50 ft) TL, 7 tonnes
FOSSIL REMAINS Minority of skeleton.
ANATOMICAL CHARACTERISTICS Insufficient information.
AGE Late Cretaceous, Turonian or Coniacian.
DISTRIBUTION AND FORMATION/S Eastern China; Jinhua.
NOTES May not be a titanosaurid.

Malarguesaurus florenciae
—— Adult size uncertain
FOSSIL REMAINS Minority of large juvenile skeleton.
ANATOMICAL CHARACTERISTICS Insufficient information.
AGE Late Cretaceous, late Turonian.
DISTRIBUTION AND FORMATION/S Western Argentina; Portezuelo.
HABITAT Well-watered woodlands with short dry season.

Tapuiasaurus macedoi
—— Adult size uncertain
FOSSIL REMAINS Nearly complete skull and minority of skeleton.
ANATOMICAL CHARACTERISTICS Snout broad and rounded.
AGE Early Cretaceous, Aptian.
DISTRIBUTION AND FORMATION/S Southeastern Brazil; Quiricó.

Malawisaurus dixeyi
—— Adult size uncertain
FOSSIL REMAINS Minority of skeleton, possibly immature.
ANATOMICAL CHARACTERISTICS Insufficient information.
AGE Early Cretaceous, Aptian.
DISTRIBUTION AND FORMATION/S Malawi; unnamed.
NOTES Remains of following unnamed taxon had been thought to belong to this taxon.

Unnamed genus and species
—— Adult size uncertain
FOSSIL REMAINS Minority of skull and partial skeletons, possibly immature.
ANATOMICAL CHARACTERISTICS Neck not elongated, broad. Tail slender.
AGE Early Cretaceous, Aptian.
DISTRIBUTION AND FORMATION/S Malawi; unnamed.
NOTES Apparently not *Malawisaurus*. Skull not short and deep as previously restored.

Gobititan shenzhouensis
—— 20 m (65 ft) TL, 15 tonnes
FOSSIL REMAINS Minority of skeleton.
ANATOMICAL CHARACTERISTICS Insufficient information.
AGE Early Cretaceous, Albian.
DISTRIBUTION AND FORMATION/S Central China; Xinminbo.

Chubutisaurus insignis
—— 18 m (60 ft) TL, 12 tonnes
FOSSIL REMAINS Two partial skeletons.
ANATOMICAL CHARACTERISTICS Insufficient information.
AGE Early Cretaceous, Albian.
DISTRIBUTION AND FORMATION/S Southern Argentina; upper Cerro Barcino.

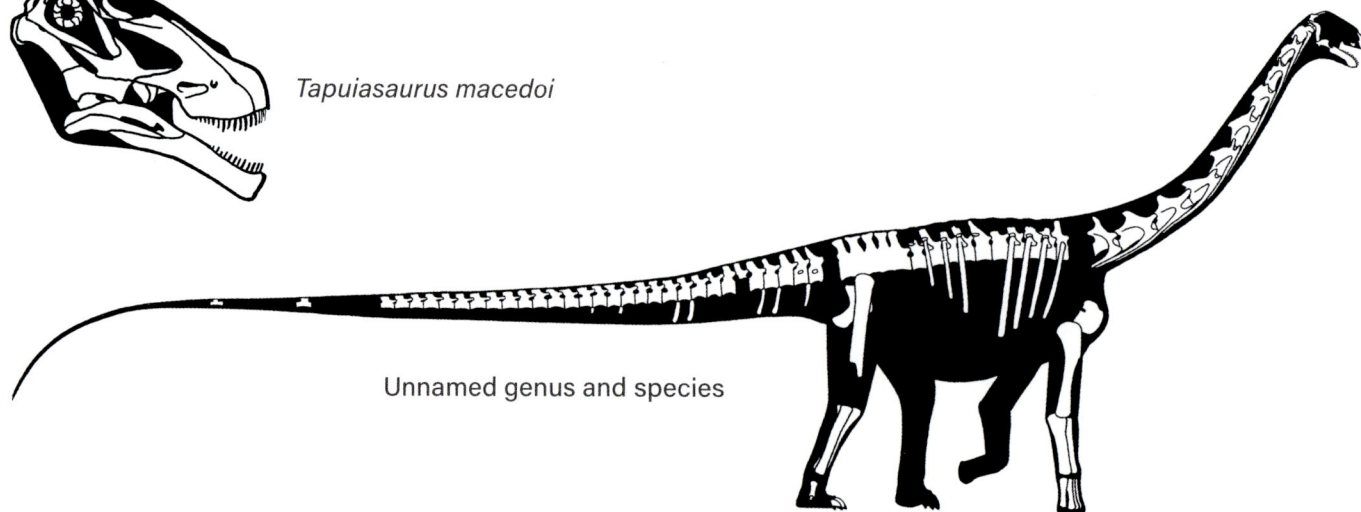

Tapuiasaurus macedoi

Unnamed genus and species

Tapuiasaurus macedoi

Austrosaurus mckillopi
—— 20 m (65 ft) TL, 15 tonnes
FOSSIL REMAINS Minority of a few skeletons.
ANATOMICAL CHARACTERISTICS Insufficient
information.
AGE Early Cretaceous, Albian.
DISTRIBUTION AND FORMATION/S Northeastern
Australia; Allaru.

Savannasaurus elliottorum
—— 15 m (50 ft) TL, 7 tonnes
FOSSIL REMAINS Partial skeleton.
ANATOMICAL CHARACTERISTICS Insufficient
information.
AGE Early Cretaceous, latest Albian.
DISTRIBUTION AND FORMATION/S Northeastern
Australia; lower Winton.
HABITAT Well-watered, cold winters.

Diamantinasaurus matildae
—— 27 m (90 ft) TL, 40 tonnes
FOSSIL REMAINS Minority of skull and four skeletons,
adult and juvenile.
ANATOMICAL CHARACTERISTICS Fingers and thumb
claw present.
AGE Late Cretaceous, late Cenomanian and/or early
Turonian.
DISTRIBUTION AND FORMATION/S Northeastern
Australia; upper Winton.
HABITAT Well-watered, cold winters.
HABITS That the fossil plant materials in the gut
contents are from conifers, angiosperms, and seed ferns
indicates a generalist diet from tree crowns down to
ground cover. Material not well chewed before
swallowing.
NOTES May include *Australotitan cooperensis*.

Andesaurus delgadoi
—— 15 m (50 ft) TL, 6 tonnes
FOSSIL REMAINS Minority of skeleton.
ANATOMICAL CHARACTERISTICS Insufficient
information.
AGE Late Cretaceous, early Cenomanian.
DISTRIBUTION AND FORMATION/S Western Argentina;
lower Candeleros.
HABITAT Short wet season, otherwise semiarid with open
floodplains and riverine forests.

Epachthosaurus sciuttoi
—— 13 m (45 ft) TL, 5 tonnes
FOSSIL REMAINS Majority of skeleton.
ANATOMICAL CHARACTERISTICS Some fingers present.
AGE Late Cretaceous, late Cenomanian or Turonian.
DISTRIBUTION AND FORMATION/S Southern Argentina;
lower Bajo Barreal.
HABITAT Seasonally wet, well-forested floodplain.

Aegyptosaurus baharijensis
—— 15 m (50 ft) TL, 6 tonnes
FOSSIL REMAINS Minority of skeleton.
ANATOMICAL CHARACTERISTICS Insufficient
information.
AGE Late Cretaceous, Cenomanian.
DISTRIBUTION AND FORMATION/S Egypt; Bahariya.
HABITAT Coastal mangroves.

Atacamatitan chilensis
—— Adult size uncertain
FOSSIL REMAINS Minority of skeleton.
ANATOMICAL CHARACTERISTICS Insufficient information.
AGE Late Cretaceous.
DISTRIBUTION AND FORMATION/S Northern Chile; Tolar.

Choconsaurus baileywillisi
—— 17 m (55 ft) TL, 10 tonnes
FOSSIL REMAINS Partial skeleton.
ANATOMICAL CHARACTERISTICS Hand elongated.
AGE Late Cretaceous, middle Cenomanian.
DISTRIBUTION AND FORMATION/S Western Argentina; lower Huincul.
HABITAT Seasonally arid open woodlands.

Gandititan cavocaudatus
—— 14 m (45 ft) TL, 5 tonnes
FOSSIL REMAINS Partial skeleton.
ANATOMICAL CHARACTERISTICS Neck moderately long.
AGE Late Cretaceous, late Cenomanian or early Turonian
DISTRIBUTION AND FORMATION/S Southeastern China; upper Zhoutian.

Baotianmansaurus henanensis
—— 20 m (65 ft) TL, 15 tonnes
FOSSIL REMAINS Minority of skeleton.
ANATOMICAL CHARACTERISTICS Insufficient information.
AGE Late Cretaceous.
DISTRIBUTION AND FORMATION/S Eastern China; Gaogou.
NOTES Position within titanosaurs uncertain.

COLOSSOSAURIANS/LITHOSTROTIANS

LARGE TO ENORMOUS TITANOSAURIDS OF THE LATE EARLY CRETACEOUS TO THE END OF THE DINOSAUR ERA, MOST CONTINENTS.

ANATOMICAL CHARACTERISTICS Fairly uniform. Necks short to long. Tails more flexible. Fingers and thumb claws absent. Egg tooth at tip of snout of hatchlings.
HABITS Dozens of spherical 0.15 m (6 in) eggs deposited in irregular shallow nests 1–1.5 m across (3–5 ft). Nests were probably covered with vegetation that generated heat through fermentation, or they were placed near geothermal heat sources, nests formed large nesting areas. Parents probably abandoned nests.
NOTES The relative placement and contents of the two groups are highly different. The juvenile egg tooth is likely to have been a wider dinosaur feature. The last of the sauropods.

Elaltitan lilloi
—— Adult size uncertain
FOSSIL REMAINS Minority of skeleton.
ANATOMICAL CHARACTERISTICS Insufficient information.

AGE Late Cretaceous, late Cenomanian or Turonian.
DISTRIBUTION AND FORMATION/S Southern Argentina; lower Bajo Barreal.
HABITAT Seasonally wet, well-forested floodplain.

Titanosaur hatchlings

Rinconsaurus caudamirus
—— 11 m (36 ft) TL, 2.5 tonnes
FOSSIL REMAINS Parts of several skeletons.
ANATOMICAL CHARACTERISTICS Neck moderately long.
AGE Late Cretaceous, Turonian or Coniacian.
DISTRIBUTION AND FORMATION/S Western Argentina; Rio Neuquén.
HABITAT Well-watered woodlands with short dry season.

Traukutitan eocaudata
—— 24 m (80 ft) TL, 30 tonnes
FOSSIL REMAINS Minority of skeleton.
ANATOMICAL CHARACTERISTICS Insufficient information.
AGE Late Cretaceous, middle Santonian.
DISTRIBUTION AND FORMATION/S Central Argentina; lowermost Bajo de la Carpa.

Bonitasaura salgadoi
—— 10 m (33 ft) TL, 2 tonnes
FOSSIL REMAINS Minority of skull and skeleton.
ANATOMICAL CHARACTERISTICS Behind lower teeth, a short, cutting beak appears to be present. Neck moderately long.
AGE Late Cretaceous, late Santonian.
DISTRIBUTION AND FORMATION/S Central Argentina; upper Bajo de la Carpa.
HABITS Predominantly grazed ground cover, also able to rear to high browse. Appears to have complemented the cropping ability of its front teeth with a supplementary beak immediately behind.
NOTES Apparently the only beaked sauropod yet known. Other sauropods known to have had similarly broad, square, ground-grazing beaks were rebbachisaurid diplodocoids like *Nigersaurus*.

Chucarosaurus diripienda
—— 25 m (80 ft) TL, 30 tonnes
FOSSIL REMAINS Minority of skeleton.
ANATOMICAL CHARACTERISTICS Lightly built.
AGE Late Cretaceous, middle Cenomanian? or early Turonian.
DISTRIBUTION AND FORMATION/S Western Argentina; lower Huincul.
HABITAT Seasonally arid open woodlands.

Muyelensaurus pecheni
—— 11 m (36 ft) TL, 2.5 tonnes
FOSSIL REMAINS Minority of skeleton.
ANATOMICAL CHARACTERISTICS Lightly built.
AGE Late Cretaceous, late Turonian.
DISTRIBUTION AND FORMATION/S Western Argentina; Portezuelo.
HABITAT Well-watered woodlands with short dry season.

Petrobrasaurus puestohernandezi
—— 20 m (65 ft) TL, 20 tonnes
FOSSIL REMAINS Partial skeleton.
ANATOMICAL CHARACTERISTICS Lightly built.
AGE Late Cretaceous, early or middle Santonian.
DISTRIBUTION AND FORMATION/S Southern Argentina; upper middle Plottier.

Drusilasaura deseadensis
—— 18 m (60 ft) TL, 11 tonnes
FOSSIL REMAINS Minority of skeleton.
ANATOMICAL CHARACTERISTICS Insufficient information.
AGE Late Cretaceous, Turonian.
DISTRIBUTION AND FORMATION/S Southern Argentina; upper Bajo Barreal.
HABITAT Seasonally wet, well-forested floodplain.

Mendozasaurus neguyelap
—— 20 m (65 ft) TL, 16 tonnes
FOSSIL REMAINS Minority of a few skeletons.
ANATOMICAL CHARACTERISTICS Neck fairly short. Vertebral spines very broad.
AGE Late Cretaceous, Turonian to Coniacian.
DISTRIBUTION AND FORMATION/S Western Argentina; Rio Neuquén.
HABITAT Well-watered woodlands with short dry season.

Atsinganosaurus velauciensis
—— Adult size uncertain
FOSSIL REMAINS Minority of skeleton, possibly immature.
ANATOMICAL CHARACTERISTICS Insufficient information.
AGE Late Cretaceous, late Campanian.
DISTRIBUTION AND FORMATION/S Southeastern France; Grès a Reptiles.
HABITAT Coastal.

Garrigatitan meridionalis
—— Adult size uncertain
FOSSIL REMAINS Minority of skeletons, some immature.
ANATOMICAL CHARACTERISTICS Insufficient information.
AGE Late Cretaceous, late Campanian.
DISTRIBUTION AND FORMATION/S Southeastern France; Grès a Reptiles.
HABITAT Coastal.

Ampelosaurus atacis
—— 16 m (50 ft) TL, 8 tonnes
FOSSIL REMAINS Minority of a few skeletons.
ANATOMICAL CHARACTERISTICS Teeth broad, line most of length of dentary.

AGE Late Cretaceous, late Campoanian and/or early Maastrichtian.
DISTRIBUTION AND FORMATION/S France; Grès de Labarre, middle Marnes Rouges Inférieures, Grès de Saint-Chinian.
NOTES *Ampelosaurus* shows that broad-toothed sauropods survived until the last stage of the dinosaur era.

Jainosaurus septentrionalis
—— 18 m (60 ft) TL, 11 tonnes
FOSSIL REMAINS Partial skeleton.
ANATOMICAL CHARACTERISTICS Insufficient information.
AGE Late Cretaceous, middle to late Maastrichtian.
DISTRIBUTION AND FORMATION/S Central India; Lameta.
NOTES May be same genus as *Titanosaurus indicus*, which is based on inadequate remains.

Arrudatitan maximus
—— 15 m (50 ft) TL, 6 tonnes
FOSSIL REMAINS Partial skeleton.
ANATOMICAL CHARACTERISTICS Insufficient information.
AGE Late Cretaceous, probably Campanian or Maastrichtian.
DISTRIBUTION AND FORMATION/S Southern Brazil; Adamantina.
HABITAT Semiarid.

Maxakalisaurus topai
—— 13 m (45 ft) TL, 4 tonnes
FOSSIL REMAINS Minority of skeleton.
ANATOMICAL CHARACTERISTICS Neck moderately long.
AGE Late Cretaceous, probably Campanian or Maastrichtian.
DISTRIBUTION AND FORMATION/S Southern Brazil; Adamantina.

Aeolosaurus rionegrinus
—— 14 m (45 ft) TL, 5 tonnes
FOSSIL REMAINS Minority of skeleton.
ANATOMICAL CHARACTERISTICS Insufficient information.
AGE Late Cretaceous, probably Campanian or Maastrichtian.
DISTRIBUTION AND FORMATION/S Southern Argentina; Angostura Colorada.

Bravasaurus arrierosorum
—— Adult size uncertain
FOSSIL REMAINS Minority of skeleton, juvenile.
ANATOMICAL CHARACTERISTICS Insufficient information.

AGE Late Cretaceous, Campanian or Maastrichtian.
DISTRIBUTION AND FORMATION/S Northwestern Argentina; lower Cienaga del Rio Huaco.

Punatitan coughlini
—— 15 m (50 ft) TL, 6 tonnes
FOSSIL REMAINS Minority of skeleton.
ANATOMICAL CHARACTERISTICS Insufficient information.
AGE Late Cretaceous, Campanian or Maastrichtian.
DISTRIBUTION AND FORMATION/S Northwestern Argentina; upper middle Cienaga del Rio Huaco.

Gondwanatitan faustoi
—— Adult size uncertain
FOSSIL REMAINS Partial skeleton, immature.
ANATOMICAL CHARACTERISTICS Insufficient information.
AGE Late Cretaceous, probably Campanian or Maastrichtian.
DISTRIBUTION AND FORMATION/S Southern Brazil; Adamantina.

Brasilotitan nemophagus
—— 12 m (40 ft) TL, 3 tonnes
FOSSIL REMAINS Minority of skull and skeleton.
ANATOMICAL CHARACTERISTICS Insufficient information.
AGE Late Cretaceous, probably Campanian or Maastrichtian.
DISTRIBUTION AND FORMATION/S Southern Brazil; Adamantina.

Adamantisaurus mezzalirai
—— 13 m (43 ft) TL, 4 tonnes
FOSSIL REMAINS Minority of skeleton.
ANATOMICAL CHARACTERISTICS Insufficient information.
AGE Late Cretaceous, probably Campanian or Maastrichtian.
DISTRIBUTION AND FORMATION/S Southern Brazil; Adamantina.

Lirainosaurus astibiae
—— Adult size uncertain
FOSSIL REMAINS Minority of several skeletons, possibly immature.
ANATOMICAL CHARACTERISTICS Insufficient information.
AGE Late Cretaceous, late Campanian and/or early Maastrichtian.
DISTRIBUTION AND FORMATION/S Northern Spain; Marnes Rouges Inferieures.
HABITAT Large island.

Lohuecotitan pandafilandi

—— 12 m (40 ft) TL, 3 tonnes

FOSSIL REMAINS Partial skull and skeleton.
ANATOMICAL CHARACTERISTICS Insufficient information.
AGE Late Cretaceous, late Campanian and/or early Maastrichtian.
DISTRIBUTION AND FORMATION/S Central Spain; upper Villabla de la Sierra.
HABITAT Coast of large island.

Bustingorrytitan shiva

—— 30 m (100 ft) TL, 50 tonnes

FOSSIL REMAINS Minority of skull and skeleton.
ANATOMICAL CHARACTERISTICS Not robustly built. Hand elongated.
AGE Late Cretaceous, middle Cenomanian.
DISTRIBUTION AND FORMATION/S Western Argentina; lower Huincul.
HABITAT Seasonally arid open woodlands.
NOTES In same size class as *Argentinosaurus*, *Patagotitan*, unnamed genus *giganteus*, *Puertasaurus*.

Paralititan stromeri

—— 27 m (90 ft) TL, 40 tonnes

FOSSIL REMAINS Minority of skeleton.
ANATOMICAL CHARACTERISTICS Insufficient information.
AGE Late Cretaceous, Cenomanian.
DISTRIBUTION AND FORMATION/S Egypt; Bahariya.
HABITAT Coastal mangroves.
NOTES Early claims that *Paralititan* rivaled the largest titanosaurs in size were incorrect.

Pellegrinisaurus powelli

—— 25 m (80 ft) TL, 30 tonnes

FOSSIL REMAINS Minority of skeleton.
ANATOMICAL CHARACTERISTICS Insufficient information.
AGE Late Cretaceous, early or middle Campanian.
DISTRIBUTION AND FORMATION/S Central Argentina; upper Anacleto or lower Allen.
NOTES A problematically large number of Anacleto and Allen titanosaurs have been named, smaller examples including *Neuquensaurus*

australis, *Barrosasaurus casamiquelai*, *Narambuenatitan palomoi*, *Pitekunsaurus macayai*, *Rocasaurus muniozi*, *Bonatitan reigi*, and *Overosaurus paradasorum* may have been the young of other titanosaurs from this formation. At least one of the known species may have laid the numerous eggs found in this formation.

Ibirania parva

—— 6 m (20 ft), 500 kg (1,000 lb)

FOSSIL REMAINS Minority of skeleton.
ANATOMICAL CHARACTERISTICS Insufficient information.
AGE Late Cretaceous, late Santonian or early Campanian.
DISTRIBUTION AND FORMATION/S Southern Brazil; lower? Sao Jose Rio Preto.
NOTES Apparantly one of the smallest sauropods.

Antarctosaurus wichmannianus

—— 17 m (55 ft) TL, 10 tonnes

FOSSIL REMAINS Minority of skull? And skeleton.
ANATOMICAL CHARACTERISTICS Front of snout broad and squared off.
AGE Late Cretaceous, late Santonian or early Campanian.
DISTRIBUTION AND FORMATION/S Western Argentina; Anacleto, level uncertain.
HABITAT Seasonally semiarid.
HABITS Jaws adapted to browse swaths of plant material, perhaps at ground level.

Rapetosaurus krausei

—— Adult size uncertain

FOSSIL REMAINS Majority of skulls and a skeleton, large juvenile.
ANATOMICAL CHARACTERISTICS Snout broad and rounded, head hourglass-shaped in top view. Neck long.

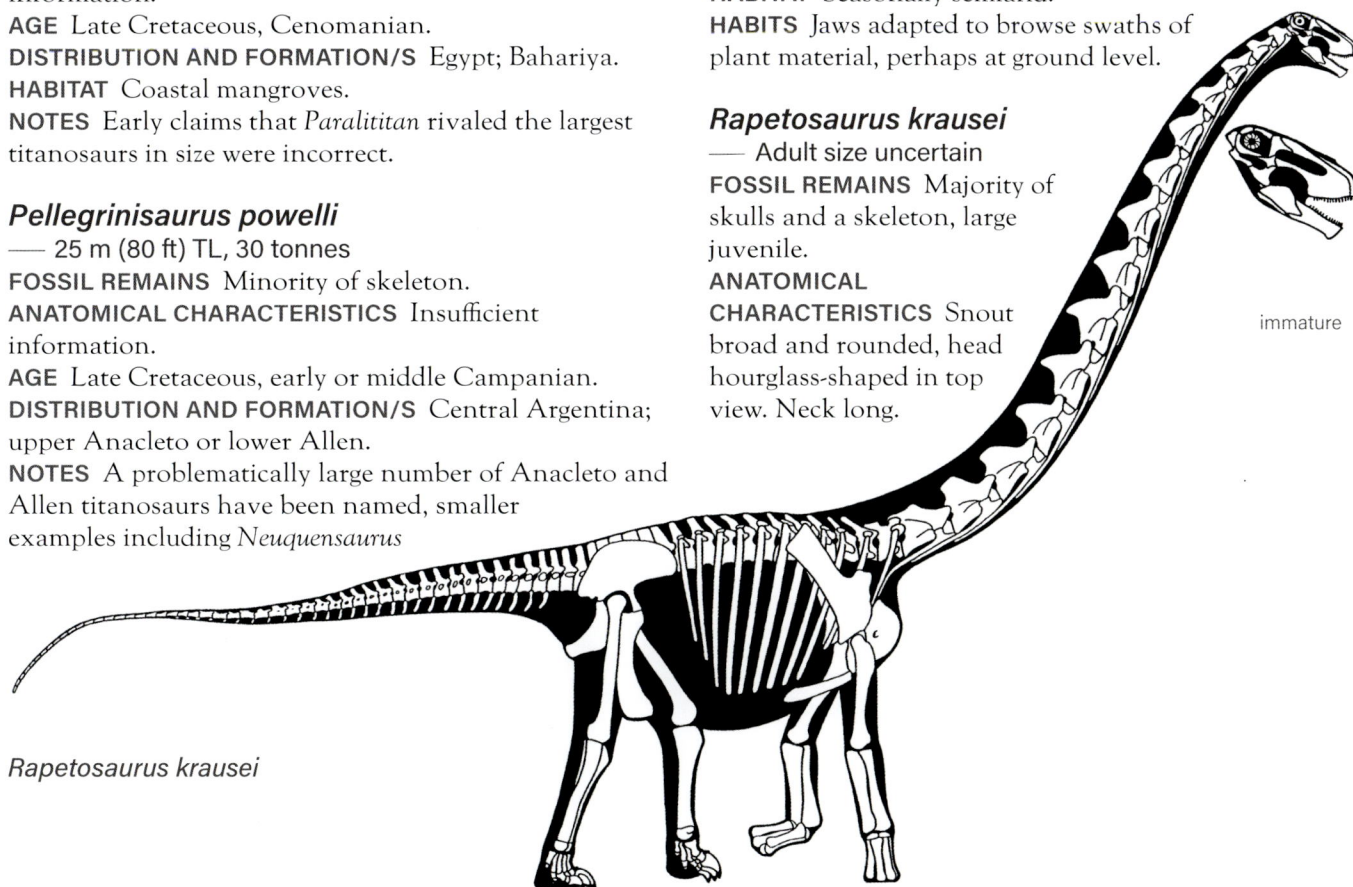

immature

Rapetosaurus krausei

AGE Late Cretaceous, Campanian.
DISTRIBUTION AND FORMATION/S Madagascar; Maevarano.
HABITAT Seasonally dry floodplain with coastal swamps and marshes.
HABITS High-level browser.
NOTES Herbivorous ornithischians apparently absent from habitat.

Unnamed genus *giganteus*
—— 30 m (100 ft) TL, 55 tonnes
FOSSIL REMAINS Minority of skeleton.
ANATOMICAL CHARACTERISTICS Arm very long, so shoulders high. Limbs elongated.
AGE Late Cretaceous, Turonian or Coniacian.
DISTRIBUTION AND FORMATION/S Western Argentina; Rio Neuquén.
HABITAT Well-watered woodlands with short dry season.
NOTES Originally placed in *Antarctosaurus*. In same size class as *Argentinosaurus*, *Patagotitan*, *Bustingorrytitan*, *Puertasaurus*.

Futalognkosaurus dukei
—— 24 m (80 ft) TL, 29.5 tonnes
FOSSIL REMAINS Majority of skeleton(s).
ANATOMICAL CHARACTERISTICS Neck long.
AGE Late Cretaceous, late Turonian.
DISTRIBUTION AND FORMATION/S Western Argentina; Portezuelo.
HABITAT Well-watered woodlands with short dry season.
HABITS Probable high-level browser.

NOTES Is not entirely certain if one or two skeletons are known, so proportions of skeletal restoration not certain, not as large as previously estimated.

Patagotitan mayorum
—— 31 m (100 ft) TL, 57 tonnes
FOSSIL REMAINS Some partial skeletons.
ANATOMICAL CHARACTERISTICS Neck long. Shoulders moderately high.
AGE Early Cretaceous, latest Albian.
DISTRIBUTION AND FORMATION/S Southern Argentina; middle Cerro Barcino.
HABITAT Seasonally very dry.
NOTES Some mass estimates have been excessive. Largest land animal represented by most of skeleton, these remains have allowed the size of other titanosaurus to be better restored and estimated.

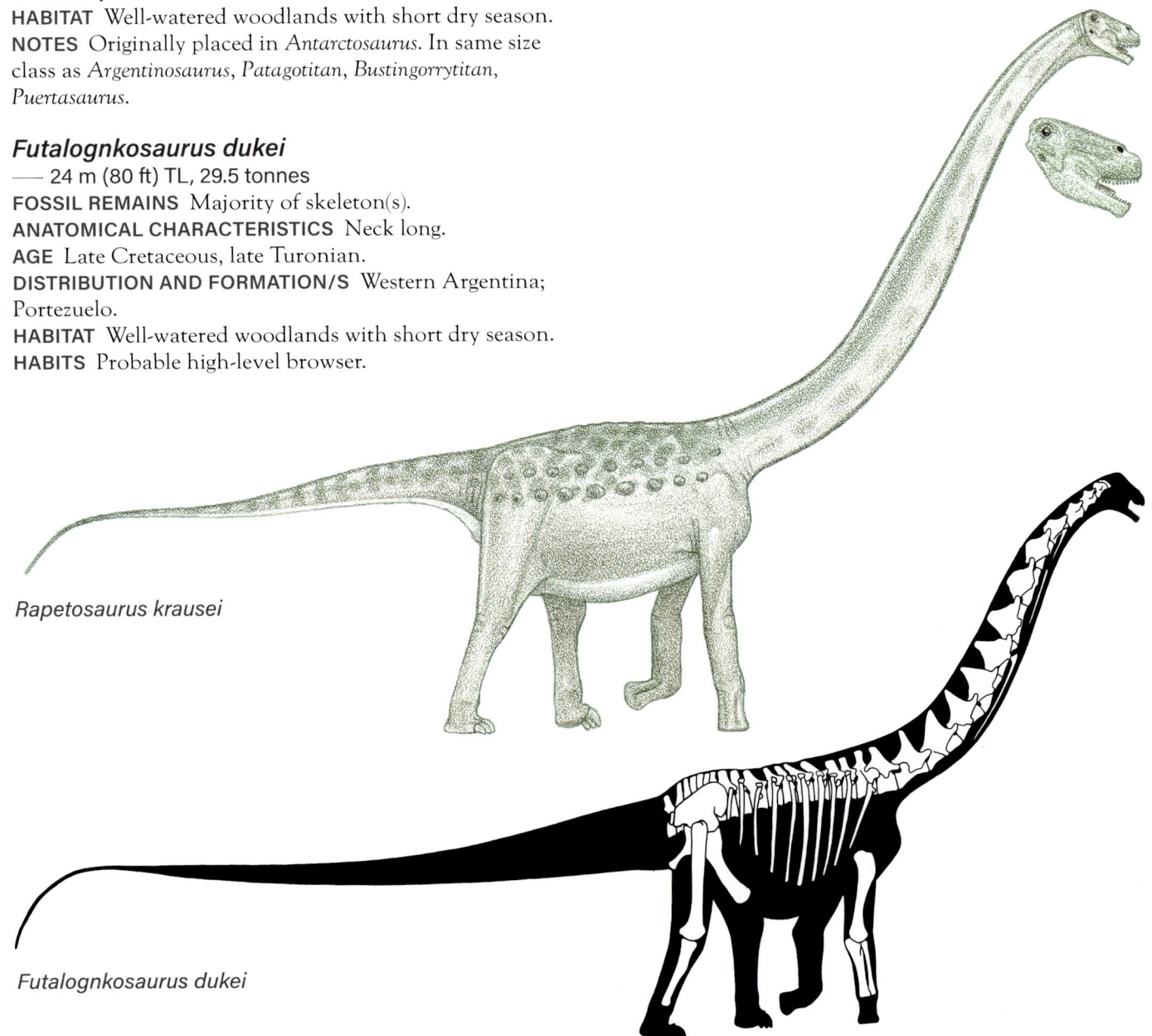

Rapetosaurus krausei

Futalognkosaurus dukei

Supertitanosaur

Patagotitan mayorum

Argentinosaurus huinculensis

—— 35 m (115 ft) TL, 80 tonnes
FOSSIL REMAINS Minority of skeleton and some other elements.
ANATOMICAL CHARACTERISTICS Insufficient information.
AGE Late Cretaceous, late Cenomanian.
DISTRIBUTION AND FORMATION/S Western Argentina; lowermost upper Huincul.
HABITAT Seasonally arid open woodlands.
NOTES Largest land animal known from substantial remains.

Dreadnoughtus schrani

—— 25 m (82 ft) TL, 33.4 tonnes
FOSSIL REMAINS Majority of skeleton.
ANATOMICAL CHARACTERISTICS Standard for group.
AGE Late Cretaceous, Campanian and/or Maastrichtian.
DISTRIBUTION AND FORMATION/S Southern Argentina; Cerro Fortaleza.
NOTES Initial suggestions of exceptional size exaggerated.

Notocolossus gonzalezparejasi

—— 27 m (90 ft) TL, 40 tonnes
FOSSIL REMAINS Minority of skeleton.
ANATOMICAL CHARACTERISTICS Foot short and stout.
AGE Late Cretaceous, late Coniacian or early Santonian.
DISTRIBUTION AND FORMATION/S Southern Argentina; lowermost Plottier.

Puertasaurus reuili

—— 30 m (100 ft) TL, 65 tonnes
FOSSIL REMAINS Minority of skeleton.
ANATOMICAL CHARACTERISTICS Neck moderately long.
AGE Late Cretaceous, early Maastrichtian.

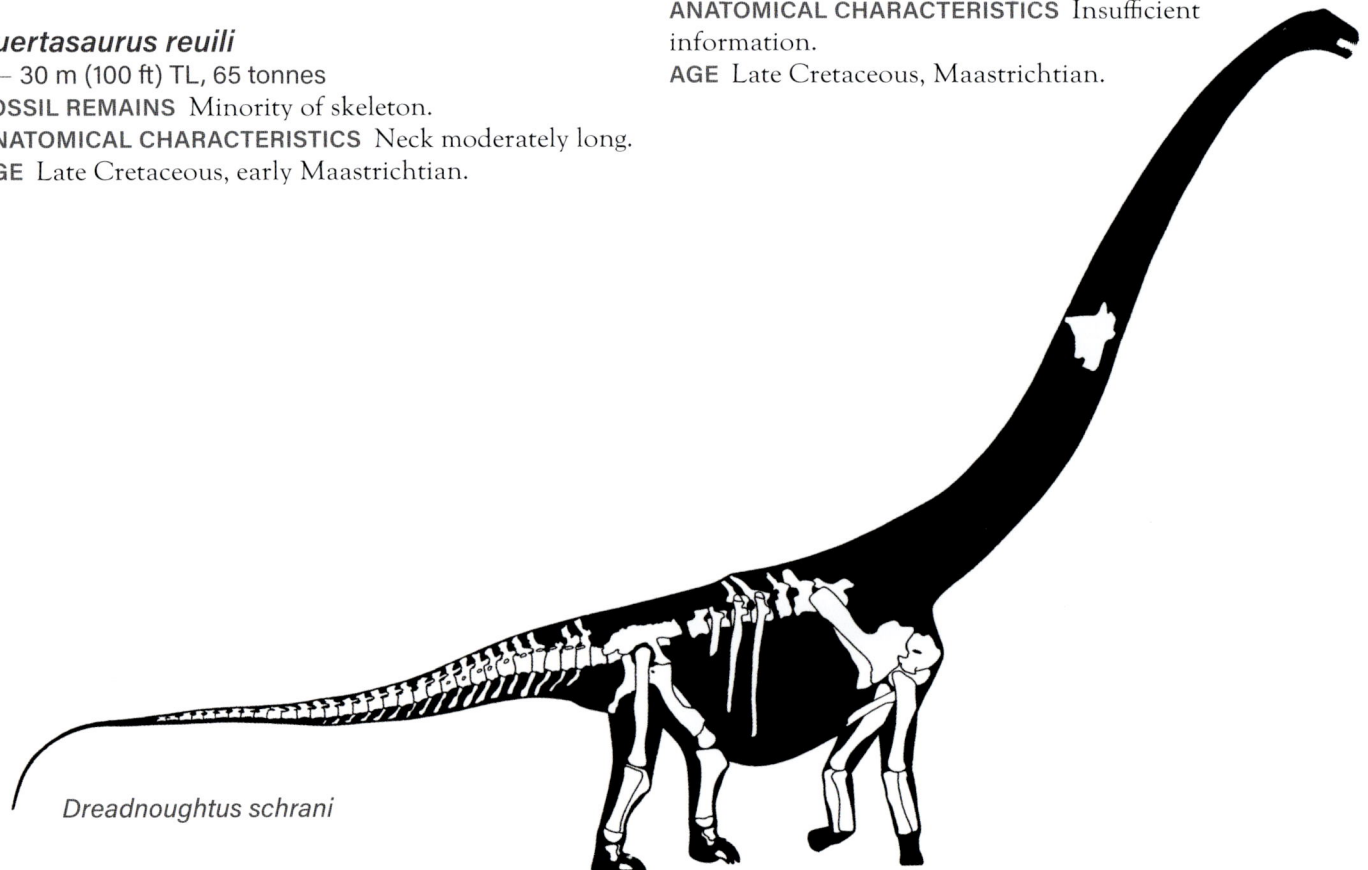

DISTRIBUTION AND FORMATION/S Southern Argentina; Pari Aike.
HABITAT Short wet season, otherwise semiarid with open floodplains and riverine forests.
NOTES Indicates that supersized sauropods survived until the final stage of the dinosaur era.

Sonidosaurus saihangaobiensis

—— Adult size uncertain
FOSSIL REMAINS Minority of skeleton, possibly immature.
ANATOMICAL CHARACTERISTICS Insufficient information.
AGE Probably late Late Cretaceous.
DISTRIBUTION AND FORMATION/S Northern China; Iren Dabasu.
HABITAT Seasonally wet-dry woodlands.

Abditosaurus kuehnei

—— 15 m (50 ft) TL, 6 tonnes
FOSSIL REMAINS Partial skeleton.
ANATOMICAL CHARACTERISTICS Neck moderately long.
AGE Late Cretaceous, early Maastrichtian.
DISTRIBUTION AND FORMATION/S Northeastern Spain; lowermost Conques.
HABITAT Large island.

Jiangxititan ganzhouensis

—— Adult size uncertain
FOSSIL REMAINS Minority of skeleton.
ANATOMICAL CHARACTERISTICS Insufficient information.
AGE Late Cretaceous, Maastrichtian.

Dreadnoughtus schrani

DISTRIBUTION AND FORMATION/S
Southeastern China; Nanxiong.
NOTES Presence in Maastrichtian indicates it was a titanosaurid.

Uberabatitan ribeiroi

—— Adult size uncertain
FOSSIL REMAINS Partial juvenile skeleton.
ANATOMICAL CHARACTERISTICS Neck moderately long.
AGE Late Cretaceous, late Maastrichtian.
DISTRIBUTION AND FORMATION/S
Southeastern Brazil; upper Marilia.

Baurutitan britoi

—— 13 m (43 ft) TL, 4 tonnes
FOSSIL REMAINS Three partial skeletons, at least some immature.
ANATOMICAL CHARACTERISTICS Neck long.
AGE Late Cretaceous, late Maastrichtian.
DISTRIBUTION AND FORMATION/S Southeastern Brazil; lower Serra da Galga.
NOTES Probably includes *Trigonosaurus pricei*.

Alamosaurus sanjuanensis?

—— 25 m (80 ft) TL, 30 tonnes
FOSSIL REMAINS Possibly a few bones of several individuals.

ANATOMICAL CHARACTERISTICS Insufficient information.
AGE Late Cretaceous, early Maastrichtian.
DISTRIBUTION AND FORMATION/S New Mexico; lowermost Ojo Alamo.
NOTES Original fossil is one bone, and others are also not highly distinctive, so status of taxon problematic. A substantial number of partial fossils assigned to this taxon from the southwestern United States from the late Campanian to late Maastrichtian are very unlikely to represent one species or even genus. These titanosaurs represent a reinhabitation of southwestern North America by sauropods from South America or, less probably, Asia after a sauropod hiatus over most of the Late Cretaceous.

Utetitan zellaguymondeweyae

—— 20 m (65 ft) TL, 20 tonnes
FOSSIL REMAINS Partial skeleton, other possible remains, immature to juvenile.
ANATOMICAL CHARACTERISTICS Neck moderately long.
AGE Late Cretaceous, late Maastrichtian.
DISTRIBUTION AND FORMATION/S Utah, Texas?; lower North Horn, lower Black Peaks?, uppermost Javelina?
NOTES Usually and incorrectly placed in much earlier *Alamosaurus*. Some or all contemporary fossils from Texas may or may not belong to taxon.

juvenile

adult

Utetitan zellaguymondeweyae

SALTASAURS

MEDIUM-SIZED LITHOSTROTIANS OF THE LATE CRETACEOUS OF EURASIA AND SOUTH AMERICA.

ANATOMICAL CHARACTERISTICS Uniform. Necks short by sauropod standards.

Isisaurus colberti
—— 18 m (60 ft) TL, 15 tonnes
FOSSIL REMAINS Partial skeleton.
ANATOMICAL CHARACTERISTICS Neck moderately long, very deep. Humerus very long.
AGE Late Cretaceous, Maastrichtian.
DISTRIBUTION AND FORMATION/S Central India; Lameta.
NOTES While upper arm is very long, relative fore and aft limb proportions are not known.

Mansourasaurus shahinae
—— 8 m (25 ft) TL, 1.5 tonnes
FOSSIL REMAINS Minority of skeleton.
ANATOMICAL CHARACTERISTICS Insufficient information.
AGE Late Cretaceous, late Campanian.
DISTRIBUTION AND FORMATION/S Egypt; upper Quseir.

Magyarosaurus dacus or Paludititan nalatzensis
—— 6 m (20 ft), 500 kg (1,000 lb)
FOSSIL REMAINS A dozen partial skeletons.
ANATOMICAL CHARACTERISTICS Standard for group.
AGE Late Cretaceous, late Maastrichtian.
DISTRIBUTION AND FORMATION/S Romania; Sânpetru.
HABITAT Forested island.
NOTES Whether the remains represent one or more taxa, and which has priority, await further research. Small size of most individuals implies island dwarfism, but some researchers cite larger sauropod fossils and higher estimate of size of island as contrary evidence.

Titanomachya gimenezi
—— 9 m (30 ft) TL, 3 tonnes
FOSSIL REMAINS Minority of skeleton.
ANATOMICAL CHARACTERISTICS Insufficient information.
AGE Late Cretaceous, Maastrichtian.
DISTRIBUTION AND FORMATION/S Southern Argentina; middle La Colonia.
HABITAT Wet and dry seasons with open floodplains and riverine forests.
NOTES Exceptionally small for a continental sauropod, causal reasons are obscure in part due to incomplete remains.

Udelartitan celeste
—— 15 m (50 ft) TL, 12 tonnes
FOSSIL REMAINS Minority of two skeletons, one immature.
ANATOMICAL CHARACTERISTICS Insufficient information.
AGE Early Late Cretaceous.
DISTRIBUTION AND FORMATION/S Western Uruguay; Guichón.
HABITAT Semiarid.

Laplatasaurus araukanicus
—— Adult size uncertain
FOSSIL REMAINS Minority of skeleton, possibly immature.
ANATOMICAL CHARACTERISTICS Insufficient information.
AGE Late Cretaceous, late Campanian.
DISTRIBUTION AND FORMATION/S Central Argentina; Allen.
HABITAT Semiarid coastline.
NOTES Many specimens have been removed from taxon, relationships with and naming of some Allen Formation titanosaurs, including *Saltasaurus robustus* and *Panamericansaurus schroederi*, are poorly understood,

Saltasaurus (= Neuquensaurus) australis
—— 7.5 m (24 ft) TL, 1.8 tonnes
FOSSIL REMAINS Partial skeletons.
ANATOMICAL CHARACTERISTICS Standard for group.
AGE Late Cretaceous, early Campanian.
DISTRIBUTION AND FORMATION/S Western Argentina; upper Anacleto.
HABITAT Seasonally semiarid.
NOTES May include *Microcoelus patagonicus* juveniles.

Saltasaurus robustus
—— 8 m (25 ft) TL, 2 tonnes
FOSSIL REMAINS A few partial skeletons.
ANATOMICAL CHARACTERISTICS Standard for group.
AGE Late Cretaceous, late Campanian.
DISTRIBUTION AND FORMATION/S Central Argentina; Allen.
HABITAT Semiarid coastal.
NOTES May be direct ancestor of *S. loricatus*.

Saltasaurus loricatus

—— 8.5 m (27 ft) TL, 2.6 tonnes

FOSSIL REMAINS Minority of skull and half a dozen partial skeletons.

ANATOMICAL CHARACTERISTICS Standard for group.

AGE Late Cretaceous, probably early Maastrichtian.

DISTRIBUTION AND FORMATION/S Northern Argentina; Lecho.

Nemegtosaurus (= Quaesitosaurus) orientalis

—— Adult size uncertain

FOSSIL REMAINS Partial skull.

ANATOMICAL CHARACTERISTICS Head standard for group.

AGE Late Cretaceous, latest Campanian.

DISTRIBUTION AND FORMATION/S Mongolia; Barun Goyot.

HABITAT Semidesert with some dunes and oases.

HABITS Fed on vegetation along streams and oases.

NOTES Shows that this sauropod dwelled in a desert, as do some elephants. May be direct ancestor of *N. mongoliensis*.

Nemegtosaurus mongoliensis
(= Opisthocoelicaudia skarzynskii) (see p. 168)

—— 13+ m (43+ ft) TL, 9.5 tonnes1

FOSSIL REMAINS A nearly complete skull and minority of skeleton, and majority of skeleton.

ANATOMICAL CHARACTERISTICS Snout broad and rounded, head strongly midconstricted, hourglass-shaped in top view. Skeleton massively constructed. Shoulders same height as hips. Hands broad.

AGE Late Cretaceous, latest Campanian and/or earliest Maastrichtian.

DISTRIBUTION AND FORMATION/S Mongolia; lower Nemegt.

HABITAT Temperate, well-watered, dense woodlands with seasonal rains and winter snow.

NOTES *Nemegtosaurus* and *Opisthocoelicaudia* are often considered different sauropods, but both are titanosaurs with only minor differences from the same level of the Nemegt, so are probably same genus if not species, any variations may be sex-related.

Nemegtosaurus (= Quaesitosaurus) orientalis

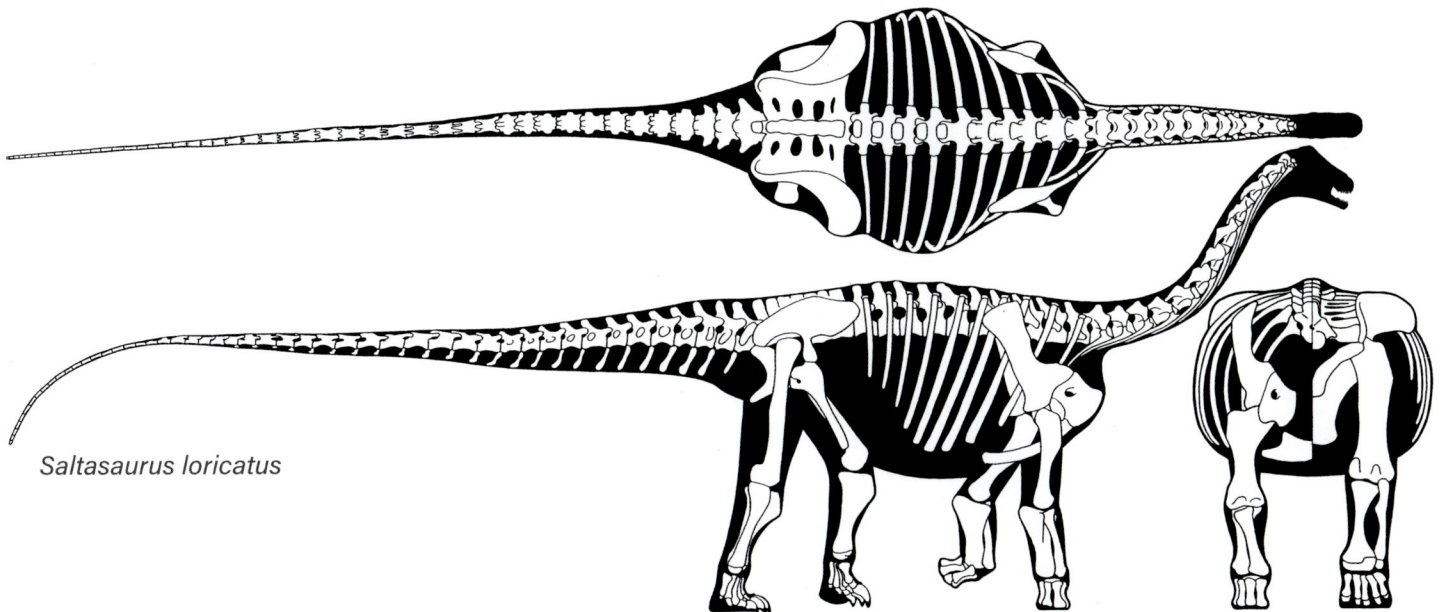

Saltasaurus loricatus

Nemegtosaurus mongoliensis
(= Opisthocoelicaudia skarzynskii)

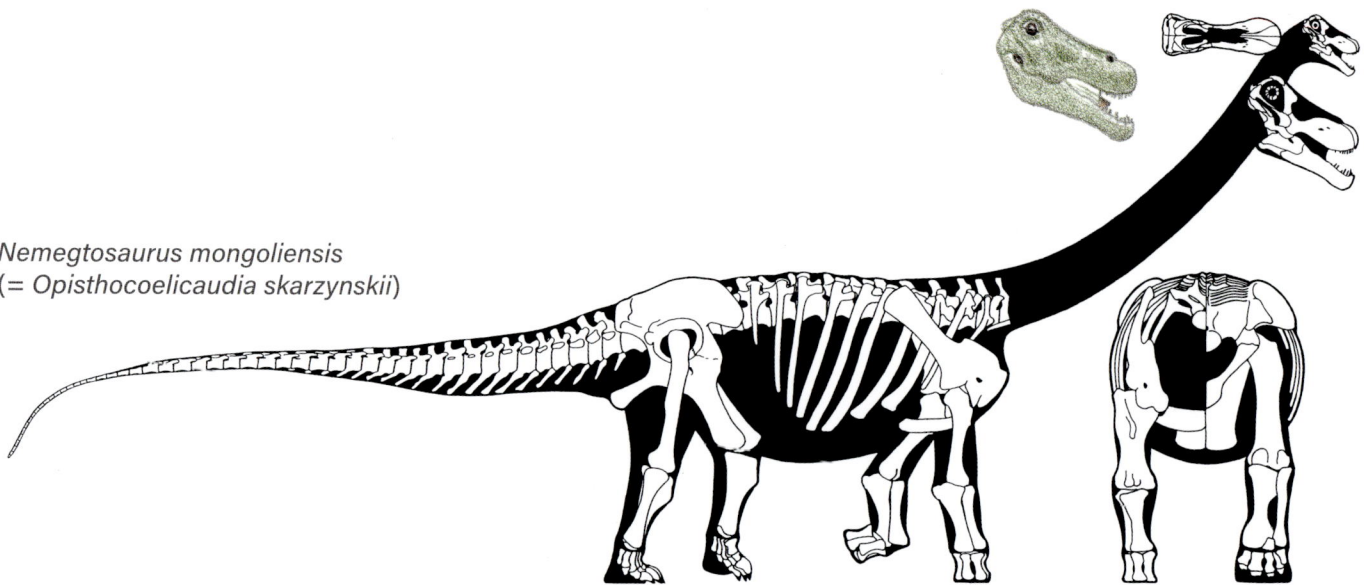

Nemegtosaurus mongoliensis

ADDITIONAL READING

Brett-Surman, M., and J. Farlow. 2011. *The Complete Dinosaur*. 2nd ed. Bloomington: Indiana University Press.

Hallett, M., and M. Wedel. 2016. *The Sauropod Dinosaurs: Life in the Age of Giants*. Baltimore: Johns Hopkins University Press.

Molina-Perez, R., and A. Larramendi. 2020. *Dinosaur Facts & Figures: The Sauropods and Other Sauropodomorphs*. Princeton: Princeton University Press.

INDEX TO SAUROPODOMORPH TAXA

INDEX TO FORMATIONS

When a formation is cited more than once on a page, the number of times is indicated in parentheses.